T0341179

MILK PRODUCTION MANAGEMENT

Milk Production Management, as the name implies, provides the information on different aspects related to Milk Production Management. The information in this book will be of practical utility for actual feeding of animals e.g. chapters on various rations, nutrient requirement tables, feeding of pregnant/lactating animals, feeding of calves, silage making, hydroponics technique, azolla production different feeds and fodders, fodder cultivation, computation of rations for dairy animals, feeding during scarcity periods etc. In this book different topics like common disease problems of dairy animals and their prevention and control, methods of selection, different breeding systems, semen collection and artificial insemination, different biotechniques used in animal husbandry, milking methods, embryo transfer technique, judging of cows and buffaloes, milk synthesis and milk secretion, record keeping at dairy farms, reproductive aspects of dairy animals etc. are also covered.

The book also covers different terms related to animal husbandry. This book is written in simple understandable language with description of those concepts which are useful for actual management of animals.

Dr. Prafullakumar V. Patil, presently working as a Farm Manager at Veterinary College, Udgir. He did his Master degree in Animal Nutrition in 2006 from MAFSU, Nagpur. He has qualified ICAR, NET examination in 2010. He has been actively engaged in research, extension and clinical activities during the last 13 years, and also engaged in teaching from the last four years. He has published 32 research papers in national and international journals along with 230 popular articles.

Dr. (Mrs.) Matsyagandha K. Patil, presently working as an Assistant Professor at Veterinary College, Udgir. She did her Master degree in Veterinary Pharmacology and Toxicology in 2005 and Ph.D. in 2016 from MAFSU, Nagpur. She has been actively engaged in teaching, research, and extension activities during the last 11 years. She has published 19 research papers in national and international journals along with more than 100 popular articles.

MILK PRODUCTION MANAGEMENT

P.V. PATIL
M.K. PATIL

College of Veterinary and Animal Sciences
Kaulkhed Road,
Udgir, Dist. Latur, Udgir-413 517 (Maharashtra)

Maharashtra Animal & Fishery Sciences, University
Futala Lake Road, Seminary Hills,
Nagpur-440 001 (Maharashtra)

NARENDRA PUBLISHING HOUSE
DELHI (INDIA)

First published 2021
by CRC Press
2 Park Square, Milton Park, Abingdon, Oxon, OX14 4RN

and by CRC Press
6000 Broken Sound Parkway NW, Suite 300, Boca Raton, FL 33487-2742

CRC Press is an imprint of Informa UK Limited

British Library Cataloguing-in-Publication Data
A catalogue record for this book is available from the British Library

Library of Congress Cataloging-in-Publication Data
A catalog record has been requested

ISBN: 978-0-367-62737-9 (hbk)
ISBN: 978-1-003-11055-2 (ebk)

CONTENTS

Preface .... *vii*

1. Introduction to Animal Husbandry 1
2. Distinguishing Characteristics of Indian and Exotic Breeds of Dairy Animals and their Performance 9
3. External Body Parts of Cows and Buffaloes 20
4. Feed Nutrients Required by Animal Body 23
5. Feeding Standards 31
6. General Dairy Farm Practices- Identification, Dehorning, Castration, Exercising, Grooming, Weighing. 35
7. Systems of Housing Dairy Animals and Maintenance of Hygiene and Sanitation at Dairy Farm Premises 48
8. Care of Animals at Calving and Feeding and Management of Calves 58
9. Management of Lactating and Dry Cows and Buffaloes 68
10. Identification Common Feeds and Fodder or Classification of Feedstuffs .. 72
11. Preparation of Rations for Adult Animals 76
12. Measures of Feed Energy 86
13. Systems of Breeding of Dairy Animals 89
14. Dairy Farm Records and their Maintenance 94
15. Common Disease Problems of Dairy Animals, their Prevention and Control 100
16. Digestive System of Cattle/Buffaloes 116
17. Male Reproductive System 118

18. Female Reproductive System 120
19. Estrous Cycle in Cows / Buffaloes 121
20. Artificial Insemination and Its Advantages 124
21. Nutrients and Fertility in Animals 126
22. Ovulation, Fertilization, Gestation, Pregnancy Diagnosis and Parturition . 129
23. Methods of Selection of Dairy Animals 135
24. Structure and Function of Mammary System 143
25. Milk Secretion and Milk Let Down 146
26. Methods of Milking, Procedure of Milking and Practices for Quality Milk Production 149
27. Factors Affecting Milk Composition of Animals 159
28. Cleaning and Sanitation of Milking Equipments 163
29. Embryo Transfer and their Role in Animal Improvement 167
30. Introduction to Biotechniques in Dairy Animal Production 172
31. Demonstration of Semen Collection, Processing 178
33. Handling and Restraining of Dairy Animals 181
33. Judging of Cows and Buffaloes 186
34. Preparation of Animal for Show 189
35. Silage Making 191
36. Nutritional Management and Milk Composition 194
37. Improved Varieties of Fodders for Animals 197
38. Hydroponics Technique for Fodder Production 20s3
39. Azolla Production and Its use in Animal Feeding 206
40. Feeding Care of Animals During Scarcity Period 208
Questions 213
References 225

PREFACE

This book contains important terms related to animal husbandry, objectives, various tables related to Animal Nutrition, differences. Tables content different chemical compositions, nutrient requirements of different animals, formulations of ration. In this book topics on managemental practices required for milk production are mainly covered along with clean milk production, sanitation, silage making, azolla production, hydroponics technique and fodder production.

This book is specially prepared as readymade notes for Dairy technology and dairy management students. It gives idea for practical feeding of animals. Along with therotical topics most of the practical aspects are covered under text matter so as to the book become more practical importance.

Thanks are due to parents, brother Pravin and Sister Kranti and childs (Ranu and Sourav), without whose co-operation this book could not have been brought out. I am also thankful to Jaya publishing house, New Delhi for kind co-operation.

Dr. P.V. Patil

Dr. (Mrs.) M.K. Patil

CHAPTER 1

INTRODUCTION TO ANIMAL HUSBANDRY

Animal Husbandry:-It is the branch of agriculture concerned with the breeding of farm animals: includes

- Cattle
- Pigs
- Sheep
- Horses
- Poultry

Domestication of wild animals started in the Prehistoric period, and they were:

- Sheep (northern Iraq)
- Goats (same region)
- Historical evidence shows that Veterinary Science was developed during Vedic Era in India and livestock used to play an important role in the society during 3000 B.C, as evidenced from Mohanjadaro and Harappa Civilization.
- The importance and role of livestock gradually increased and during 2000 B.C, Veterinary profession was a flourishing practice which can be traced from **'Atharvaveda' and 'Rigveda'**.
- Aryans who settled around riverine Northern India around 2400 – 1500 B.C were solely dependent on agriculture and livestock.
- Cattle were most prized possessions and were symbol of wealth and status.
- During the rule of Ashoka (300 B.C), hundreds of well equipped hospitals were established and veterinary profession gained much more importance.

- Vishnupuranam and Matysapuran described the criteria for selection of bulls for breeding purpose.
- Primitive man first used the members of family bovidae as a source of food. Domestication began when these animals were used as draft animals.
- Milking qualities were just sufficient for rearing of young ones.
- As civilization developed, feed became more abundant, methods of caring livestock improved. Under man's selection they acquired qualities like rapid growth, better fat storage in body and increased milk production.

Modern Developments in Animal husbandry

- Selective breeding
- Advances in animal nutrition
- Vet medicine
- Artificial insemination
- Embryo transfer

Farm Specialists

- Breeders
- Milkers
- Feeders
- Health specialists Vets

The Cattle Family includes

- Cow
- Milking cow
- Heifer
- Milker
- Calf
- Ox
- Steer
- Bull

Cows' Life

- Gestation period is 9 months 9 days or 280 ± 5 days
- Newborn calf weighs 35-40 kg in crossbred cow & 20-25 kg in indigenous cow
- Dairy cows provide 90% of world's milk production
- Farmers use machines to milk 100 cows for 1 hour
- Cattle and buffalo life span= 18-22 years

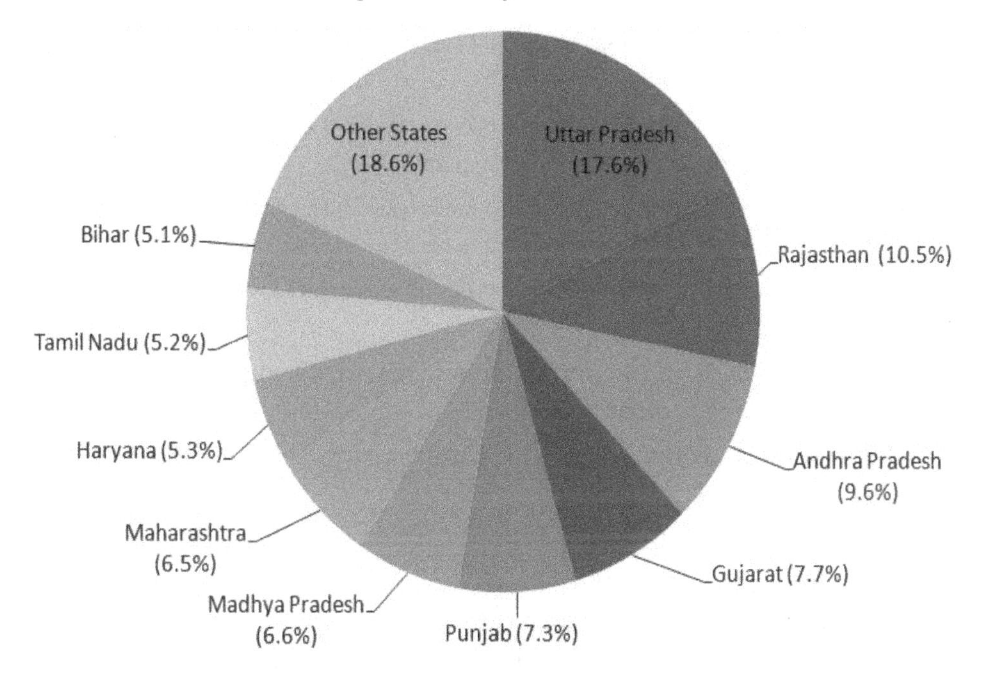

Contribution of State in Milk Production in India

Rank of India in total milk production: First

Total milk production in 2017-18= 176.33 million tonnes.

Cow milk production in India = 76 million tonnes

Buffalo milk production in India=100.33 million tonnes

Uttar pradesh has highest buffalo population in India

Uttar Pradesh is largest milk producer state in India (27770 thousand tonnes)

- Rajasthan is the second largest milk producing state.

- Gujrat is third largest milk producer in India.
- Bihar and Jharkhand have largest cattle population in India.
- Total livestock population is 535.78 million in India and shown increase of 4.6% over livestock Census-2012.
- The Bovine population (Cattle, buffalo, Mithun and Yak) is 302.79 million in 2019 which shows an increase of 1.0% over the previous census.
- The total cattle population in India is 192.49 million in 2019 showing an increase of 0.8% over previous census.
- The female cattle (cow) population is 145.12 million, increased by 18.0% over the previous census (2012).
- The exotic /crossbred and indigenous/ non-descript cattle population in India is 50.42 million and 142.11 million respectively.
- The indigenous/ non-descript female cattle population has increased by 10% in 2019 as compared to previous census.
- The population of the total exotic/crossbred cattle has increased by 26.9% in 2019 as compared to previous census.
- Buffalo population in India =109.85 millions showing an increase of about 1.0% over previous census (2012).
- The total milch animal (in milk and dry) in cows and buffaloes is 125.34 million, an increase of 6.0% over the previous census.
- The total sheep population in India is 74.26 million in 2019, increased by 14.1% over previous census.
- The goat population in the country in 2019 is 148.88 million showing an increase of 10.1% over the previous census.
- The total pig population in India is 9.06 million in 2019, declined by12.03% over the 2012 census.
- The total Mithun population in India is 3.9 lakhs in 2019, increased by 30% over previous census.
- The Yak population in India is Fifty eight thousand in 2019, decreased by 24.67% over previous census.
- The total horses and ponies in India are 3.4 lakhs in 2019, decreased by 45.6% than the previous census.
- The total population of Mules in the India is Eighty four thousand in 2019, decreased by 57.1% than previous census.
- The population of Donkeys in India is 1.2 lakhs in 2019, decreased by 61.23% than previous census.

- The Total Camel population in India is 2.5 kakhs in 2019, decreased by 37.1% than previous census.
- The total poultry in India is 851.81 Million in 2019, increased by 16.8% than previous census.
- The total Backyard poultry in India is 317.07 Million in 2019, increased by 45.8% than previous census.
- The total commercial poultry in India is 534.74 Million in 2019, increased by 4.5% than previous census.

Common Animal Husbandry Terms

- **Buller or Nymphomaniac:** A cow apparently always in heat.
- **Back crossing:** Mating of crossbred back to one of the pure parents used to produce it.
- **Balanced ration:** Ration that contains all the nutrients in right proportions and quantities as per requirement of particular animal is called balanced ration.
- **Bull Calf:** A male calf under one year of age.
- **Bull:** It is un -castrated sexually matured male of the species.
- **Bullock:** Castrated male Ox.
- **Calf starter:** Concentrate feed offered to the young calves after 2 weeks of age.
- **Calf:** A young animal of bovine species under one year of age.
- **Casting:** It is throwing down the animal slowly and safely and securing the limbs for various purposes like surgical operations, castration, hoof trimming, shearing etc.
- **Castration:** It is the removal of testicles or dysfunctioning of testicles in male.
- **Challenge feeding:** The practice of feeding higher levels of concentrate to challenge the cow to reach her maximum milk production.
- **Concentrates:** Feeds that contain less than 18% crude fibre and more than 60% TDN are called concentrates such as grains, oilcakes, grain by products etc.
- **Cow:** It is a female of bovine species that has calved at least once.
- **Cross breeding:** A system of breeding between two established breeds.

- **Cryptorchid:** A male animal in which one or both the testicles fail to descend into the scrotal sac.
- **Culling:** Removal of undesirable or unproductive animals from herd.
- **Deticking:** Removal of the external parasites like ticks, lice, mites present on the body surface of animal.
- **Deworming:** Removal of the internal gastro intestinal parasites from the body.
- **Disbudding:** Removal of the horn buds of the calf by mechanical or chemical methods to arrest growth of horns.
- **Dry period:** The time interval between date of drying off the cow to the date of next calving.
- **Energy feeds:** Feeds containing less than 20% crude protein are called energy feeds.
- **Lactation period:** The number of days a cow secretes milk following each parturition.
- **Free martin:** When twin calves of different sexes are born, the bull calf is normal whereas the heifer calf is sterile. The sterile heifer calf is called freemartin.
- **Gestation period:** The period of pregnancy in animals.
- **Grading up:** Systems of breeding in which pure bulls are used for improvement in non descript females for several generations.
- **Heifer calf:** A female calf under one year of age.
- **Heifer:** A female individual above 1 year of age that has not yet calved.
- **Inbreeding:** A system of breeding between very closely related animals.
- **Inheritance:** Transmission of genes from parents to the offspring in next generation.
- **Intercalving period:** No. of days between two successive calvings.
- **Lactation Curve:** The graphical representation of the rate of milk secretion during lactation is called Lactation Curve.
- **Lactation length:** The time interval between the date of calving to the date of drying the animal expressed in days.
- **Maintenance ration:** A ration given daily to the animal to maintain in resting non production condition with good health.
- **Open animal:** Female animals that have not been bred.
- **Parturition:** Act of delivery in animals.

- **Pasture:** Fodder crops grown on the land for grazing animals.
- **Pedigree Bull:** The bull whose ancestral record is known.
- **Persistency:** Ability of the animal to sustain good daily milk for a longer period i.e, the slope of descending phase of lactation curve is known as Persistency.
- **Phenotype:** The visible character of an individual animal.
- **Production ration:** A portion of the ration given daily in excess of maintenance requirement for purpose of growth, production and work.
- **Protein supplements:** Feeds that contain 20% or more protein are called protein supplements.
- **Ration:** The total amount of feed offered to an animal is during a 24 hour period of time is called ration.
- **Roughage:** Feeds that contain more than 18% crude fiber and less than 60% TDN are called roughage such as hay, silage, fodder etc.
- **Scrub Bull:** It is non-descript type of stray village cattle.
- **Selection:** The process of including certain animals in a population for becoming parents of next generation.
- **Breed:** Animal having a common origin and characteristics that distinguish them from other groups within the same species.
- **Service period:** The period between parturition to successful conception expressed in days.
- **Silage making:** It is a method of conservation of green fodder in which Controlled fermentation of green fodder in a specially prepared silo is carried out.
- **Stud Bull:** Bull that is used for breeding purposes.
- **Test cross:** Mating of a crossbred back to its recessive parent.
- **Dry cow:** A cow that is not producing milk.
- **Variation:** It is a tool to measure differences of character or trait between animals.
- **Weaning:** Separation of the calf from the cow and feeding them artificially.
- **Animal husbandry:** It is defined as the branch of science and art of management of common farm practices like scientific feeding, breeding, health care of common domestic animals aiming for maximum returns
- **Puberty:** The period of life at which the reproductive organs first became functional. This is characterised by estrus and ovulation in the female and semen production in the male.

- **Agalactia:** Failure to secrete milk following parturition.
- **Anorexia:** Lack of appetite.
- **Artificial insemination:** The injection of mechanically procured semen into the reproductive tract of the female without coition and with the aid of mechanical or surgical instuments.
- **Colostrum:** The first milk produced by the female immediately after giving birth to young.
- **Conception:** The action of conceiving or becoming pregnant.
- **Drying off:** The act of causing a cow to cease lactation in preparation for her next lactation.
- **Estrus cycle:** The period from one estrus to next.
- **Estrus:** The period of sexual excitement in the female.
- **Milk let down:** The squeezing of milk out of the udder tissue into the gland and teat cisterns.
- **Milk vein:** A large blood vessel which runs under the center of the animals belly towards the udder.
- **Reproductive cycle:** The sexual cycle of the non-pregnant female characterised by occurance of estrus at regular intervals.
- **Teaser:** A male made incapable by vasectomy or by use of an apron to prevent copulation of impregnating a female.
- **Crisscrossing or Rotational crossing:** Mating of a hybrid to three established breeds in a rotational manner.
- **Silage:** is a fermented, high-moisture stored fodder which can be fed to cattle, buffalo etc.

CHAPTER 2

DISTINGUISHING CHARACTERISTICS OF INDIAN AND EXOTIC BREEDS OF DAIRY ANIMALS AND THEIR PERFORMANCE

CATTLE BREEDS

Breed

A group of animals related by descent and similar in most characters like general appearance, features, size, configuration are said to be breed.

Species

A group of individuals which have certain common characteristics that distinguish them from other groups of individuals.

1. RED SINDHI

- Origin: Around Karachi and Hyderabad of Pakistan
- Colur: Deep dark red, Body –medium sized, compact, proportionate, head is moderate.
- Horns:- thick, stumpy with blunt points, upwards
- Milk yield/lactation: 1750 kg
- Fat percentage: 4.6 %

CLASSIFICATION OF CATTLE BREEDS

Milch	Dual	Draft	Exotic
• Sahiwal	• Deoni	• Khillar	• Holstein Friesian
• Red sindhi	• Gaolao	• Amritmahal	• Jersey
• Tharparkar	• Kankrej	• Hallikar	• Brown Swiss
• Gir	• Haryana	• Nagori	• Red Dane
	• Rathi	• Bargur	
	• Ongole	• Bachaur	
	• Nimari	• Malvi	
	• Dangi	• Khergarh	
	• Mewati	• Kangayam	
		• Siri	
		• Krishna Valley	

2. THARPARKAR

- Origin: Tharparkar district in Pakistan, Amarkot, Navkot district of Gujrat and Jodhapur district of Rajasthan
- Characters: Body: medium sized, strongly built, forehead flat.
- Ears: slightly pendulous with rich yellow colour
- Colour: White grey with light grey strip along with back bones.
- Milk yield: 1136-1200 kg/lactation
- Age of first calving: 43 to 56 month
- Intercalving period: 15 to 16 months

3. HARIANA

- Origin: Rohtak, Hissar, Karnal district of Haryana state
- Characters: Head: carried high and forehead is slightly convex
- Ears: small and active
- Horns: thin, short carrying inward and upward
- Nevel flap: seen in females
- Milk yield: 1000-1150 kg/lactation
- Fat %= 4.5 to 4.6 %
- Age at first calving = 58 months & ICP- 15 to 20 months

4. GIR

- Synoname: Kathiawari,Surti
- Origin: Gir forest and hilly area of South Kathiawar in Gujrat.
- Characters: Head: massive, forehead: prominent & extremely bulging
- Horns: big, curved, turned backward
- Ears: long, pendulous, curved with notch at the tip.
- Eyes: prominent, almond shape appearance, sleepy look.
- Colour: Varies from red with light patches of white and red
- Tail: long, reaching to fetlck joint
- Udder: Capacious. Milk yield: 1590 kg/lactation. Fat: 4.5%

5. SAHIWAL

- Synoname:Lola, Multani
- Origin: Montgomery district in Pakistan and in Punjab.
- Body: heavy with short legs
- Skin: loose
- Head: broad with stumpy horns. Loose horns in females
- Colour: red brown with or without white splashes.
- Dewlap: Voluminous & pendulous sheeth
- Tail-long reaching to ground
- Udder: Capacious. Milk yield: 2250 kg/lactation
- Fat: 5 to 5.2 %

6. RATHI

- Origin: Alwar & Rajputana of Rajasthan, around Bikaner
- Colour: White or light grey
- Well built and deep chest.
- Face: straight
- Eyes: wide and large
- Ears: short & pendulous
- Tail: short with black switch
- Bullocks are powerfull and active, suitable for field & road work

7. DANGI

- Origin: Ahmednagar, Nasik, Bansda, Sonkhed region of Maharashtra and Gujrat
- Colour: Broken red & white or black & white
- Body: medium sized
- Skin: oily
- Head: small, projecting forehead
- Horns: short & thick
- Ears: small
- Hardy breed, bullocks are excellent workers in rice field & heavy rainfall tracks

8. DEONI

- Synoname: Dongerpatti
- Origin: Devani, Udgir tahsils of Udgir, North West and western portion of Hyderabad
- Characters: Gir like
- Forehead: pronounced
- Horns: outward and backward curve
- Face: lean but not clean cut
- Colour: Black & white or red and white irregular patches or spots.
- Chest: deep, heavy dewlap, pendulous sheath Milk yield: 900 kg/lactation

9. GAOLAO

- Origin: Southern Madhya Pradesh, Maharashtra
- Colour: females are pre white; males are grey over head, neck and humps
- Face: very long, narrow with flat foreheads
- Eyes: almond shaped
- Horns: short stumpy
- Dewlap voluminous.
- Tail: comparatively short
- Milk yield: 816.5 kg/lactation

10. KANKREJ

- Origin: north Gujrat and distributed in Rann of Kutch
- One of the heavy breed, broad chest, forehead dished in centre. Horns: strong curved accompanied by skin up to some length, straight back, humps well developed, tough skin, Colour: male- silver grey, iron grey or black. In females colour markings are lighter.
- Gait of kankrej is called "Swai chal" (smooth movements of body and head noticeably high).
- Kankrej cattle are highly prized as draught cattle.
- Milk yield: 1333 kg/lactation

11. ONGOLE

- Synoname: Nellore
- Origin: Ongole tract of Andhra Pradesh, comprising Guntur, Narasaraopet, Venukonda, Kandukur talukas of Nellore & Guntur districts
- Body: Large, heavy and muscular
- Forehead: broad with stumpy horns thick at the base and firm without cracks.
- Hump: well developed, erect and filled from both sides,
- Colour: White, males are dark grey at extremities.
- Milk yield: 1255 to 2268 kg/lactation
- Major beef breed for foreign countries

12. KHILLAR

- Origin: Solapur,Sangli & Satara district of Maharashtra.

Four types of Khillar

1. Atpadi Mahal of Hanam Khillar from the southern Maharashtra
2. The Mhaswad Khillar from Solapur & Satara districts
3. The Thillari or Tapi Khillar from west Khandesh
4. The Nakali i.e. imitation Khillar

Characters of Khillar

1. Body: compact with clean-cut features.
2. This breed resembles with Amritmahal breed
3. Colour: Mhaswad Khillar-greyish males being darker over the fore-quarters & Hind-quarters with typical mottled markings on the face.

4. The Tapi khillar is White with a pink nose & hoofs
5. The Nakali Khillar is grey with tawny or brick dust colour over the forequarters
6. Head: massive, eyes & ears -small,
7. Horns-long, pointed,
8. Shoulders are tightly muscled
9. Barrel: long, compact with no loose skin
10. Hoofs are black. Fast & powerful draught breed of Maharashtra

13. RED (LAL) KANDHARI

- Origin: Kandhar, Mukhed, Biloli, Degloor tahsils of Nanded district and the borders of Andrapradesh
- Characters:
- Head: broad, forehead buldging slightly
- Ears : drooping sideways & long
- Colour: light red to brick red
- Eyes: black
- Hoofs black
- Milk yield: 530 kg/lactation
- Age at first calving-42-52 months
- ICP- 15 to 17 months

14. AMRITMAHAL

- Origin: Karanataka state
- Body: compact with short straight back, well arched ribs,
- Narrow face and prominent forehead with furrow in the middle
- Horns: long sweeping
- Hump & Dewlap-well developed
- Colour: grey body with dark head, neck,hump and quarters
- One of the best draught breeds of India

15. NAGORI

- Origin: Jodhpur and Nagore district of Rajasthan
- Supposed to be evolved from Hariana & Kankrej breeds

- Colour- generally white or grey
- Body: long, deep, powerful with straight back
- Forehead- well developed, flat
- Ears: large, pendulous
- Face: narrow & long
- It is the most useful draught breed of India

16. HALLIKAR

- Origin: Hassan & Tumkur districts of Karnataka state.
- Characters:
- Medium sized animal
- Head-long with bulging forehead,furrowed in middle
- Horns: long, graceful sweep on each side of the neck
- Face: long, Ears-small
- Colour:grey to dark grey with deep shadings on the fore and hind quarters
- Bullocks are excellent draught type suitable both for road & field work.

17. JERSEY

- Origin: Island of Jersey in the English channel
- Colour: Fawn with or without white markings
- Milk yield: 4000 liters/lactation(305 days)
- Age at first calving-38 months
- Head: have double dish
- Straight top lines, leveled rumps and sharp withers.
- Can withstand tropical and humid climate more than Holstein

18. HOLSTEIN FRIESIAN

- Origin: Holland
- Colour: Black-white
- Milk yield: 6150 liters/lactation
- Age at first calving-36 months
- Animals ruggedly built & possess large feeding capacities and udder.
- Head-long, narrow and straight

19. BROWN SWISS

- Origin: Switzerland
- Colour: Distinctly brown
- Milk yield: 5250 liters/lactation
- Head- large,dished and thick loose skin
- This breed is more heat-tolerant than Jersey
- The breed is used for ploughing & pulling carts as well as milk & beef production

They are also classified based on physical features as below:

1) Group-I : Lyre horned grey cattle with wide forehead, flat/dished face, lyre shaped horns.

   Ex: Kankrej, Hissar, Khenkalha, Malvi, Tharparkar
2) Group-II: White (or) light grey cattle with Coffin shaped skull

   Ex: Ongole, Gaolao, Hariana, Krishna valley, Mewati, Nagori, Ratti, Bachaur
3) Group-III : Animals with heavy build & curled horns

   Ex: Most milch breeds – Gir, Deoni, Red sindhi, Sahiwal, Dangi
4) Group-IV: Mysore breeds with prominent forehead & long horns

   Ex: Hallikar, Alambadi, Amritmahal, Bargur, Killari, Kangayam
5) Group-V (Hill Type): Small black, red colored with large patches of white found in hill tract in North India

   Ex: Siri in Darjeeling and Ponwar in UP
6) Group VI: Medium sized draught breeds, tight naval flap and dewlap, red and white color

   Ex: Dhani breed of Pakistan.

BUFFALO BREEDS

In India total **13** buffalo breeds are recognized. Three types of buffaloes present 1. Water buffalo, 2. Cape buffalo 3. Swamp buffalo

CLASSIFICATION OF BUFFALO BREEDS

On the basis of regions the buffalo breeds are as follows

1. Murrah group:
   i) Murrah

   ii) Nili-ravi
   iii) Kundi
   iv) Godavari
2. Gujrat group:
   i) Surti
   ii) Jaffarbadi
   iii) Mehsana
3. Uttar Pradesh:
   i) Bhadwari
   ii) Tarai
4. Central India:
   i) Nagpuri
   ii) Pandharpuri
   iii) Manda
   iv) Jarangi
   v) Khalandi
   vi) Sambalpur
5. South India:
   i) Toda
   ii) South Kanara

1. Murrah Buffalo

Habitate	**Rohtak, Hissar and Karnal district of Hariana, Punjab, Delhi & Western UP**		
Average Production Traits		**Phenotypic traits**	
305 days Milk Yield	1400-2000 Kg	Horns	Short, Tightly curved horns
Age at First Calving	42 months	Colour	Jet Black
Lactation Length	300 days	Size	Deep massive frame with short back, light head & neck
Calving Interval	430 days	Udder	Well developed
		Tail	Long tail with white switch reaching to fetlock

2. Jaffrabadi Buffalo

Habitate		Gir forest of Kathiavar	
Average Production Traits		**Phenotypic traits**	
305 days Milk Yield	2336 Kg	Horns	Heavy, inclined to droop on each side of the neck, turns up
Age at First Calving	52 months	Colour	Black
Lactation Length	343 days	Size	Body longer but no compact
Calving Interval	447 days	Udder & dewlap	Well developed
Forehead	Very prominent		

3. Surati buffalo

Habitate		South-western part of Gujarat, Anand, Nadiad & Baroda districts	
Average Production Traits		**Phenotypic traits**	
305 days Milk Yield	1600 Kg	Horns	Sickle shaped, fat which grow in downward, backward direction & tip upward forms hook
Age at First Calving	44-52 months	Colour	Black or brown hairs are rusty brown to silver grey
Lactation Length	350 days	Size	Medium sized, docile, have unique straight back
Calving Interval	461 days	Udder	Well developed
		Eyes	Bright, prominently round

4. Mehsana Buffalo

Cross between Murrah & Surati breed

Habitate		Mehsana town in north Gujarat, Palampur in Banskantha district & Radhanpur and Tharad in Saharkantha district	
Average Production Traits		**Phenotypic traits**	
305 days Milk Yield	1800 Kg	Horns	Horns are sickle shaped or slightly curved inside
Age at First Calving	43.0 months	Colour	Jet black to brown grey with white markings on face, legs, tip of tail
Lactation Length	352 days	Size	Medium sized, deep, low-set deep body
Calving Interval	16 months	Udder	Well developed
		Head	Resembles the Murrah with buldging eyes

5. Pandharpuri buffalo

Habitate		Southern Maharashtra	
Average Production Traits		Phenotypic traits	
305 days Milk Yield	1400 Kg	Horns	Long, touching hock bone
Age at First Calving	44.8 months	Colour	Blackish with grey
Lactation Length	350 days	Size	Medium
Calving Interval	465 days	Forehead	Long narrow face, Prominent thick neck
		Tail	Short, White switch common

6. Nagpuri buffalo

Habitate		Vidarbha region of Maharashtra	
Average Production Traits		Phenotypic traits	
305 days Milk Yield	1200 Kg	Horns	Long flat, curved backwards on sides of neck
Age at First Calving	55.8 months	Colour	Black
Lactation Length	270 days	Size	Medium with slightly deep back
Calving Interval	430 days	Forehead	Long, Cone shaped Straight nasal bone
		Tail	Short, White switch common

7. Bhadwari

- Copper colour, hairs scanty.

8. Manda

Colour is brown or grey with yellowish tufts of hair on knees & fetlocks, switch of tail is yellowish.

CHAPTER 3

EXTERNAL BODY PARTS OF COWS AND BUFFALOES

Mainly body parts of animal is divided into four parts

1. Head
2. Neck
3. Trunk
4. Tail

1. Head region

a) Muzzle: - The black portion above the upper lip. During sound health it is moist, while it is dry during sickness.

b) Horn: - They are situated in a pair. Functions of horns are i) self defence, ii) estimation of age & iii) for gressful appearance.

   In some of the animals horns are completely absent called as "polled animal".

c) Face: - The portion between the crest and mouth include following sub parts.

   i) Forehead: - The portion between the crest and line between the eyes itself. It differs from breed to breed.

   ii) Nose: - This includes central bridge starting from the central of line joining to eyes and elongated upto muzzle.

   iii) Mouth: - The opening for intake of feed & water.

   iv) Eyes: - Eyes are situated each side of the nose.

   v) Ears: Two ears are present.

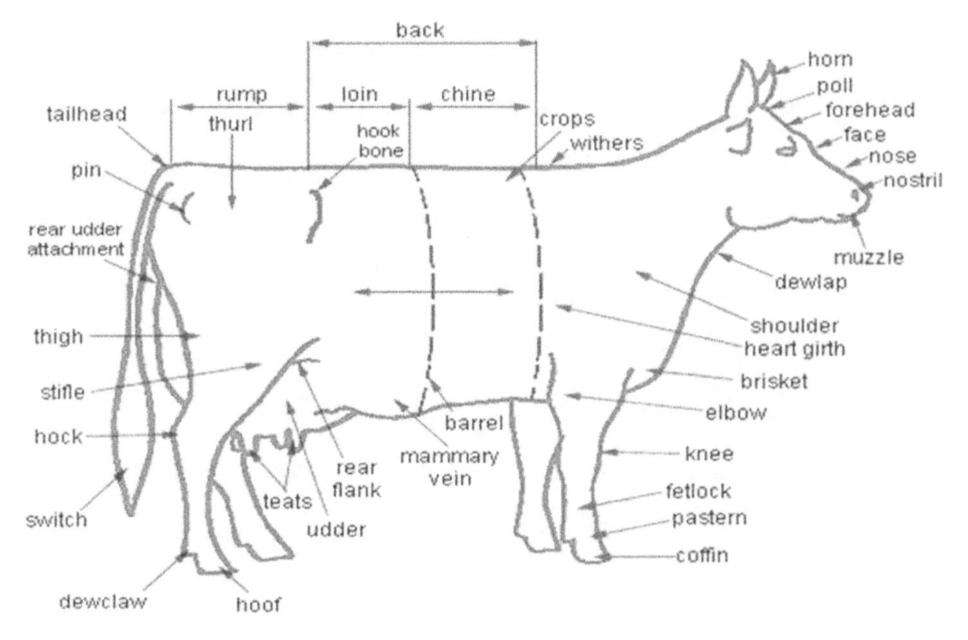

External body parts of cow

2. Neck region

Neck is the portion between the head and body. Neck region can be divided into two parts.

a) Neck crest: - It is the line between centre of hump.

b) Brisket: - Fleshy ball like structure hanging in between two forelegs.

3. Trunk region

This region is also called as body or barrel of animal. It consists of following parts.

a) Wither: - The fleshy portion below the hump and above the shoulder.

b) Back: - Portion between hump and sacrum supported by ribs. It's having parts chine, lion and hallow of flank.

c) Hip bone: - Hip bones are the raised bone of hallow of flank.

d) Chest: - Bottom portion of body covered by ribs.

e) Abdomen: - The ventral portion of body uncovered by the ribs.

f) Hump: - The buzzling and fleshy portion above the shoulder. It is well developed in male than female.

g) Arms: - Portion of leg between shoulder and portion of elbow.

h) Fore arm: - The portion between point of elbow and knee joint.

i) Shank: - The joint between knee and pastern.

j) Hoof: - The lower hardest portion of the legs is hoof. There are two digit hoofs.

k) Dewclaws: - The hars finger like projections on backside of pastern is called as dewclaws.

l) Hind quarter: - The region provides information on the development of reproductive organs, mammary gland.

m) Thigh: - It is a fleshy portion of the hindlegs in between rump and stifle joint.

n) Milk vein: - A prominent zig-zag vein starts from the heart to udder.

o) Udder: Udder is known as mammary gland. It is attached to the body in between two hindlegs and the naval. The complete udder is divided into four parts called as quarters. Each quarter extended with the tube like structure known as teat.

4. Tail region

A long whipe like structure in continuation of vertebral column. The proximal end attacched to sacrum is called as root of tail or base of tail. Middle part of the tail is as body of tail and distal part of the tail is covered with hairy bunch on the tip of the tail known as switch or switch of tail.

Anus: Extreme end of the elimentary canal located below the base of the tail.

Vulval lips: Present below the anus in females.

CHAPTER 4

FEED NUTRIENTS REQUIRED BY ANIMAL BODY

Nutrient

Any constituent of a feed that goes to produce heat and energy and to control body processes.

OR

Any substance that nourishes the body called as nutrient

Nutrients required by the animal body

1. Protein
2. Carbohydrates,
3. Fat,
4. Minerals
5. Vitamins &
6. Water

1. Protein

Complex organic compound containing C, O, N, H and also P & S.

Amino acids are present in polypeptide chain.

— **Chemical composition of protein**

Carbon = 50%

Oxygen = 23%

Nitrogen = 16%

Hydrogen = 7%

Phosphorus = 0.3 %

Sulphur = 0.3%

Characterisitics of different proteins

1. **Globular proteins**: (Enzymes, antigens and protein hormones)
   i) Albumins: Water soluble and heat coagulable.
   ii) Globulins: Insoluble in water, soluble in dilute salt solution of sod. Chloride; heat coagulable.
   iii) Glutelins: Soluble in dilute solutions of acids and alkalis, insoluble in water and dilute salt solutions.
   iv) Prolamins: Insoluble in water & dilute salt solutions; soluble in 70% ethyl alcohol.
   v) Histones: Soluble in water, non heat coagulable.
   vi) Protamins: Basic in nature, rich in arginine and found more in sperms.

2. **Conjugated proteins:**

   It consist of amino acid+ non-protein group
   i) Glycoproteins: Contain one or more heterosaccharides or their derivatives as prosthetic group; found in mucosal cells
   ii) Lipoproteins: Contain phospholipids, soluble in water; found in cell
   iii) Chromoproteins: Coloured proteins contain pigments as prosthetic group: haemoglobins
   iv) Nucleoproteins: Contain nucleic acids (DNA,RNA) in combination with protein molecules.

3. **Fibrous Proteins:**

   Insoluble, long filamentaous, resistant to digestive enzymes.
   i) Collagens: Insoluble in water. Present in skeleton or connective tissue
   ii) Keratin: Found in hair, feather, hoofs, nails, beaks and horns. Insoluble in water and less digestible
   iii) Elastin: Found in tendons and arteries, not soluble in water, dilute acids or bases.

— **Sources of protein:**

Leguminous fodder, animal byproducts, oil seed cakes etc.

Functions of proteins:

1. Protein makes up 17 to 20 % of the animal body and 3.6 % of the milk.
2. Makes up new tissues and muscles in the body.
3. Repairs the loss of body tissues of muscles.
4. Helps in growth and development of body.
5. Helps in production of enzymes & hormones.
6. Helps in production of wool, egg and meat.
7. Excess protein provides energy.

2. Carbohydrates

Organic material containing chemically CHO, which includes starch, sugars, celluloses and gums.

Formed in plants by photosynthesis and used as ready energy source in feed.

Sources of Carbohydrates in animal body:

a) **Crude fibre:** Fibrous, woody, hard material

   It contains cellulose, hemicellulose, lignin and pentosans, not easily digested in animal body.

   Rumen micro-organisms convert cellulose into simple sugars and fatty acids

   Mature fodder, straw and bhusa, dry grass contains the crude fibre.

b) **Nitrogen free extract:**

   These includes starch and sugars (soluble carbohydates). These are helpful in making up milk sugar (lactose).

Functions of Carbohydrates:

1. Provide enery more readily than other oraganic constituent.
2. Maintains body temparature.
3. Excess of carbohydates stored as fats, a reserved source of energy.
4. Important part of blood as a blood sugar.
5. Stored as glycogen in body tissues and liver.
6. Induce sweetness to milk.
7. Helps in absorption of calcium & phosphorus in growing calves.
8. Helpful in growth & multiplication of micro-organisms in the rumen.

3. Fats (Ether extract)/lipids

— **Classification of lipids:**

Lipids are divided into 1.Simple lipids, 2. Compound or conjugated lipids & Derived lipids.

1. Simple lipids are fats/waxes(Esters of fatty acids with glycerol).
2. Compound lipids are glycolipids and phospholipids.

Derived lipids are essential oil, steroids and sterol.

— **Sources of fats:**

— Oil seeds- cotton seed, linseed, groundnut, mustard, til, cereals etc.

Functions of Fats:

1. Provides energy 2.25 times more than carbohydrates.
2. Reserve source of energy.
3. Helps in formation of milk fat.
4. Makes up approx. 20% of animal body.
5. Essential constituent of protoplasm.
6. Helps in absorption of carotene.
7. Helps in availability of vitamins A,D, E, K.
8. Provides essential fatty acids like linoleic acid, linolinic and arachidonic acids.
9. Acts as shock absorber & prevent injury to internal organs.

4. Minerals

This is an inorganic matter.

Two types of minerals: 1. Major elements (Macro elements) & 2. Minor or Micro elements

1. Major minerals: Ca, P, Mg, Na, K, Cl and Sulphur
2. Minor elements: Fe, Cu, I, Co, Zn, Mn, Fl, Se, Mo, Cr. Ni

— **Sources of minerals:**

Green fodder, cereals, cakes, brans, cereal byproducts, meat meal, bone meal etc.

Functions of minerals:

1. Makes useful & essential salts for body.
2. Helpful in digestion and absorption of nutrients.
3. Increase efficiency of absorption of oxygen in blood.

4. Helps in clotting of blood.
5. Helps in production of RBCs in blood.
6. Helpful in production of enzymes and hormones.
7. Maintains osmotic pressure in cells of muscles.
8. Helps in formation of bones, teeth and muscles.
9. Responsible for permiability of cell membranes.
10. Increase palatability of ration.
11. Increases enzyme action and thereby digestion of nutrients.

5. Vitamins

A class of organic substance required by animals in minute quantities essential for metabolism, growth and development.

Vitamins are classified into 1. Fat soluble vitamins & 2. Water soluble vitamins

1. Fat soluble vitamins- Vit. A,D,E,K
2. Water soluble vitamins- Vit.C, B-Complex

— **Sources of vitamins-**

Green fodders, hay, cereals, cakes and byproducts.

Functions of vitamins:

1. Necessary for proper functioning of different activities of body.
2. Important for good health.
3. Helps in growth, development & production.
4. Helps in digestion of nutrients.
5. Provides resistance to infections in general.
6. Helps in metabolism of energy.
7. Helps in maintaining proper vision, health of skin, formation of bones, fertility, utilization of nutrient, growth of hair, appetite.

6. Water

Sources of water for animal body:

1. By voluntary intake: through drinking water
2. By feed: Through green fodder
3. Metabolic water: Water produced due to oxidation of nutrients in body (5-10%)

Functions of water:

1. Acts as a general lubricant and cleansing agent for different parts of animal body.
2. Provide rigidity and elasticity to body cells.
3. As a solvent for absorption of nutrients & excretion of waste products.
4. Helps in maintaining the osmotic pressure of body.
5. Regulates body temperature.
6. Maintains moistness in body.
7. Transportation of absorbed nutrients to various body parts.
8. It helps in gaseous exchange in the tissues & lungs.
9. The aqueous humor helps to keep up the shape and tension of the eyeball and acts as a refractive medium for light.

Nutrient requirements for maintenance per head per day

Body weight (Kg)	DM (Kg)	DCP (Kg)	TDN (Kg)	S.E. (Kg)	Ca (g)	P (g)	Carotene (mg)	Vit.A 1000 (mg)
200	3.5	0.148	1.66	1.24	5	5	21	9
250	4.5	0.168	2.02	1.56	6	6	26	11
300	5.6	0.197	2.36	1.77	7	7	32	13
350	6.7	0.227	2.7	2.02	8	8	37	15
400	7.8	0.254	3.03	2.26	9	9	42	17
450	8.9	0.282	3.37	2.51	10	10	48	19
500	9.1	0.296	3.69	2.92	11	11	53	21
550	10.1	0.336	3.71	3.18	12	12	58	23
600	11.1	0.350	3.80	3.20	12	12	64	26

Nutrient requirements for milk production per Kilogram of milk

Fat%	DCP (Kg)	TDN (Kg)	S.E. (Kg)	Ca (g)	P (g)	Carotene (mg)	Vit.A 1000(mg)
3.0	0.040	0269	0.233	2	1.4	2.5	1.8
4.0	0.045	0.316	0.275	2	1.4	2.7	2.0
5.0	0.051	0.363	0.316	2	1.4	2.9	2.2
6.0	0.057	0.411	0.357	2	1.4	3.1	2.4
7.0	0.063	0.458	0.398	2	1.4	3.3	2.6
8.0	0.069	0.506	0.439	2	1.4	3.5	2.8
9.0	0.075	0.553	0.480	2	1.4	3.7	3.4
10.0	0.081	0.602	0.521	2	1.4	3.9	3.2

Nutrient requirements for maintenance+gestation (NRC Std.) (During last two months of pregnancy)

Body weight (Kg)	DM (Kg)	DCP (Kg)	TDN (Kg)	S.E. (Kg)	Ca (g)	P (g)	Carotene (mg)	Vit.A 1000 (mg)
350	6.4	0.315	3.6	10.1	21	16	67	27
400	7.2	0.355	4.0	11.2	23	18	76	30
450	7.9	0.400	4.4	12.3	26	20	86	34
500	8.6	0.430	4.8	13.4	29	22	95	38
550	9.3	0.465	5.2	14.4	31	24	105	42
600	10.0	0.500	5.6	15.5	34	26	114	46
650	10.6	0.530	6.0	16.2	35	28	124	50
700	11.3	0.555	6.3	17.3	39	30	133	53
750	12.0	0595	6.7	18.0	42	32	143	52
800	12.6	0.630	7.1	19.0	44	34	152	61

Nutrient requirements for growing heifers per day (NRC)

Body weight (Kg)	DM (Kg)	DCP (Kg)	TDN (Kg)	S.E. (Kg)	Ca (g)	P (g)
45	0.6	0.15	0.8	2.1	3.5	2.5
70	2.0	0.22	1.5	5.4	10	8
100	3.2	0.28	2.2	8.0	13	10
150	4.5	0.35	3.0	10.8	18	14
200	5.9	0.40	3.8	13.7	21	16
250	7.3	0.43	4.5	16.3	24	18
300	8.7	0.47	5.2	18.8	27	20
350	10.2	0.50	5.9	21.3	29	22
400	11.8	0.54	6.6	23.8	30	23
450	12.5	0.59	7.0	25.3	30	23

Nutrient requirements for mature breeding bull per day (NRC)

Body weight (Kg)	DM (Kg)	DCP (Kg)	TDN (Kg)	S.E. (Kg)	Ca (g)	P (g)	Carotene (mg)	Vit.A 1000 (mg)
400	7.5	0.250	4.2	15.4	15	10	40	16
500	8.3	0.300	4.6	16.6	20	15	53	21
600	9.6	0.345	5.4	19.5	22	17	64	26
700	10.9	0.390	6.1	22.1	25	19	75	31
800	12.0	0.430	6.7	24.2	27	21	86	36
900	13.1	0470	7.3	26.4	30	23	96	41
1000	14.1	0.505	7.9	28.6	32	25	107	46

CHAPTER 5

FEEDING STANDARDS

Feeding standards are the statements or quantitative description of one or more nutrients needed by the animal for various functions of body.

Feeding standards are of three types

1. Comparative type feeding standards
   i) Hay feeding standard
   ii) Scandinavians feeding standard
2. Digestible nutrient feeding standards
   i) Grouvens feeding standard,
   ii) Wolfs feeding standard,
   iii) Wolf Lehmann feeding standard,
   iv) Heckers feeding standard,
   v) Sevage feeding standard,
   vi) Morison feeding standard,
   vii) Indian feeding standard, NRC feeding standard
3. Productive type feeding standards
   i) Kellner feeding standard,
   ii) Armsby feeding standard,
   iii) ARC feeding standards

1. Comparative feeding standards

i) **Hay feeding standard:** This is also called as hay equivalent feeding standard. It was suggested by scientist Thaer in 1810. He compared different feeds with Medow hay as a unit. In this standard the nutritive value of feed and physiological requirements of animal were not considered. It was completely based on practical feeding experience.

ii) **Scandinavians feeding standard:** It is also called Scandinavian feed unit system. It was given by Scientist Fjord in 1854. One pound of barley was used as a unit. This standard is similar to that of hay feeding standard only the unit has been changed. According to this standard one feed unit is required for each 150 pounds of body weight and additional unit for each pound of milk produced.

2. Digestible type of feeding standards

i) **Grouvens feeding standard:** This feeding standard was based on protein, carbohydrate and fat content of the feed.

ii) **Wolffs feeding standard:** The scientist Emil Von Wolff has given this feeding standard which is based on digestible protein, digestible carbohydrate and digestible fat content of feed. This standard proposes a nutritive ratio of 1: 5.4. The drawback of this feeding standard is that it does not considered the quality & quantity of milk produced.

iii) **Wolff Lehmann feeding standard:** This is the modified form of Wolff feeding standard where in quantity of milk produced is considered where as quality of milk has been neglected.

iv) **Haeckers feeding standard:** This feeding standard was ued in America & was the first feeding standard which took into consideration the quality and quantity of milk and also the separate requirements for maintenance, reproduction and production were given in this feeding standard for the first time.

v) **Savage feeding standard:** In this feeding standard it is suggested that the nutritive ratio should not be wider than 1:6 and narrow than 1:4.5. This feeding standard also increase the protein requirement by 20% than Heckers feeding standard. In this feeding standard the quantity and quality of milk produced as well as the different milk fat levels were taken into consideration. This standard suggests that 2/3 rd of dry matter requirement of animal should be supplied through roughage and 1/3rd from the concentrates.

vi) **Morison feeding standard:** The scientist F.B. Morison first presented the standards in 1915. This standard is expressed in terms of dry matter, DCP and TDN. In 1956, it was revised and ME like terms were included. This standard was further modified and the requirements for Ca, Phosphorus and carotene were included. For Indian conditions Morison feeding standards are used with certain modifications. This is refered as Modified Morisons feeding standard or Mid Morisons feeding standard.

vii) **Indian feeding standard:** This standard is given by ICAR. It includes the requirement for cattles in terms of DCP, TDN, NE, and also for Calcium, Phosphorus and Carotene. Whereas separate standards have been given for Poultry, Sheep, Pig and goat. The nutrient requirement for Poultry is expressed in terms of Crude Protein and ME.

The Indian feeding standards for cattle are based on Morison feeding standard. A separate standard have been proposed by ISI. Which is also called as BIS (Buro of Indian Standards) for poultry.

3. Productive type feeding standards

i) **Kellner feeding standard:** The scientist Kellner proposed his feeding standard which is based on starch equivalent. The starch equivalent is defined as the amount of pure starch required to produce the same amount of fat which is produced by 100kg of feed. The S.E. is similar to N.E. of feedstuff since both expressions gives productive energy of feed.

ii) **Armsby feeding standard:** It is based on Crude protein & NE value & its use is limited as compared to other feeding standards.

iii) **ARC feeding standard:** The agricultural research counsil of Briton (U.K.) publishes the feeding standards for livestock & poultry. In this feeding standards the unit of energy requirement is SE instead of TDN, ME & NE.

Advantages of feeding standards

1. Serves as a general guide for feeding of livestock.
2. Useful for practical feeding purpose.
3. Gives idea about total feed and nutrient requirement of energy specific for physiological functions.
4. Useful in planning the experiments and interpreting the results depending upon nature and objective of the investigation.
5. Useful for calculation of total requirement of herd and thus helps in planning of feeding schedule for future.
6. Being flexible feeding standards can be modified as per demands, availability and cost of feesdstuffs.

Limitations of feeding standards

1. Can not give exact needs of individual animals.
2. Unable to indicate whether or not the animals are fed properly.
3. Can not become complete guide to feeding of animals and hence difficult to use as a rule.
4. May not be useful under the situations where palatability and physical nature of feed alters its voluntary intake and thus its requirement.
5. Environment and climatic changes can alter the nutrient metabolism and therefore these cannot be useful in all such conditions.
6. It may change according to genetic makeup.
7. No useful measure of food energy is given.
8. Factors such as biological value, amino acid composition, available minerals and vitamins etc. are not taken into account.
9. Modifications are needed according to availability of feeds and also as per economic factors.

CHAPTER 6

GENERAL DAIRY FARM PRACTICES-IDENTIFICATION, DEHORNING, CASTRATION, EXERCISING, GROOMING, WEIGHING

1. Identification of farm animals

Identification of farm animals is necessary for-

i) Distinguish the animal.

ii) It helps in identification of particular animal.

iii) It is helpful for conviance of farm management.

iv) It is important for maintaining production and health record.

v) To locate the mothers calf.

Identification of farm animals is usually done by following methods,

A) Tattooing

Tattooing is the process of numbering desire number or letter inside skin of ear with the help of tattoo forceps and then rubbing tattooing ink on it. It is most suitable for marking newborn calves and pigs. Before tattooing ear should be cleaned and desire number fixed to tattooing forcep and pressed on inner side of ear.

Procedure

i) Find out and decide the number to be tattooed.
ii) Arrange the desired numbers on the tattooing forceps.
iii) Check this number of tattoo set on a piece of thick paper.
iv) Secure the calf and hold the ear horizontally.
v) Locate the place on the inner side of ear between the large vein.
vi) Clean this numbering place with spirit to remove dirt and ear wax grease.
vii) Sterilize the numbers fixed on tattoo set by using spirit.
viii) Apply some ink on the numbers.
ix) Hold the tattooing forceps with pad outside and tattoo number inside the ear.
x) Press the handle with gentle pressure stopping at the clicking sound and hold it for a while.
xi) Open and remove the tattooing forceps.
xii) Apply ink with a swab and rub well with the thumb into holes. Release the calf.
xiii) Clean the tattoo set properly.
xiv) Precaution should be taken to avoid injury to ear vein.

Advantages of tattooing

i) It is easy and permanent method
ii) It is less painful.
iii) It does not disfigure the animal.

Disadvantages of tattooing

i) Regular cleaning of ear is required to see the tattoo number.
ii) It is less useful in animals having dark pigmented skin.
iii) Reading of number needs close inspection.
iv) Mild bleeding may takes place.

Tattooing forcep/machine

B) Ear notching

It is the method of making 'V' shape cuts/ notches at specific places along the borders of ear with the help of scissor or pincers. It is commonly used for marking buffalo calves and pigs.

Method

1. Firstly, the ear is cleaned and disinfected with the help of spirit.
2. Then 'V' shape cuts/notches are made at the border of ear by using sterilised ear punch/scissor/pincers.
3. A single notch at lower side of right ear indicates the number '1' while of left ear indicates the number '3'.
4. However a single notch at the upper side of right ear indicates the number '10', while on left ear indicates the number '30'.
5. Notches should neither be made too small to close up soon nor too larger to deform the shape of ear.

Advantage

It is a permanent method of marking pigs.

Disadvantage

It causes injury which is difficult to heal.

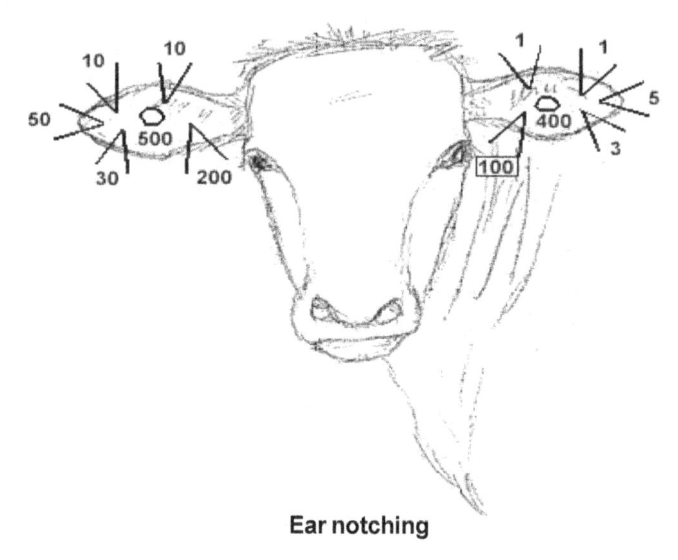

Ear notching

C) Tagging

Tagging is method of fixing tag to the ear/neck of animal. It is mostly used marking of sheep, goat and sometimes young calves.

i) **Ear tagging:** In this method, tags made of light metal or strong plastic having number stamped on it or blank tag on which desired number will be written are used. Tags are of two types i.e. self piercing and non-piercing. Prior to tagging ear is cleaned with spirit. The self piercing tag is directly pierced and locked with the help of pincers. In case of non-piercing tag, hole is made on the upper edge of ear with the help of ear notcher or ear punch and then tag is placed in the hole and fixed. While tagging, the number should be visible outside the upper edge of ear. Tag should not be too tight or too loose. Antiseptics like tincture iodine should applied at the site of tagging. Take care that tag should not be tight or too loose but leaving enough space for growth of ear.

Disadvantage: Tags usually fall off or even tear off the ear lobe.

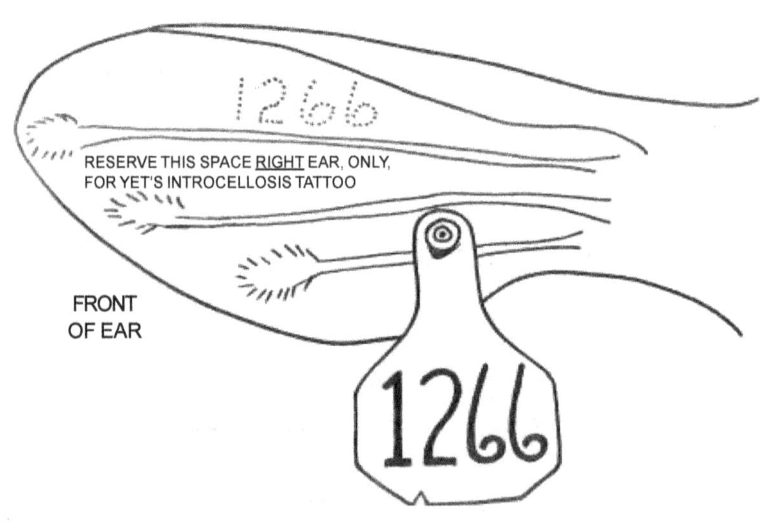

Ear tagging and tattooing

ii) **Neck tagging:** In this method metalic or plastic tag is tied in the neck chain, thread, or wire. It is temporary method of marking animals. Used when herd owners do not want to use permanent identification. Used in early age for identification to avoid injury to ears.

D) Branding

It is the method of imprinting number or any identification mark on the skin of animal by hot iron, chemical or coolent. It is mostly used for marking of cattle and buffaloes. Branding is of three types: 1. Hot branding, 2) Cold branding & 3) Chemical branding

Smaller size brands are used for branding of calves at one year of age. Branding is done when plenty of sunshine is available and prevalence of fly is less.

Branding should be done preferably on left i.e. milking side. Copper branding iron is preferred as it retains heat for longer time and hairs do not stick to it.

1) **Hot branding:** A branding iron bearing number from 0-9 or alphabetes from A-Z is used for this purpose. The required number or letter is fixed to the branding iron. It is then heated till it becomes red hot. Such red hot branding iron is applied for about 3 to 4 seconds on thigh with little pressure. Branding burns the hair and skin at the site of branding and leaves a permanent mark on the body. The fresh mark is then smeared with Zinc oxide and mustard oil to promote early healing.

Disadvantages

i) It is painful method

ii) It needs experienced person

iii) It causes permanent damage to skin and thereby reduces market price of animal.

iv) It is less useful in rainy and winter season.

2) **Chemical Branding:** In this method, branding iron is deeped in branding ink (caustic liquid or paste) and excess of ink is removed. The branding iron is then applied on skin where marking is desired. This branding causes loss of hair due to irritation to skin and a permanent scar developes at the site.

Advantages

i) It is comparatively less painful

ii) It requires minimum skill

3) **Cold branding:** This method is called as cryobranding or freeze branding.

Tie the animals hind legs. Fasten the tail to the legs. Branding site is first cleaned and soaked with alcohol for 20 to 30 seconds. Shake the branding solution well and pour it in a shallow porcelain disc or enamel. Then the branding iron is dipped in a coolent like liquid nitrogen and applied at required site of skin for short period. The coolent destroys melanocytes i.e. colour producing cells of hair. Thus hairs at branded area appear white or colourless after 50-70 days of branding.

Advantages

i) It is less painfull.

ii) It is most effective in animals with black body coat.

Comparative (parameters) merits and demerits of different methods of marking

Sr. No.	Parameters	Tattooing	Branding	Tagging	Ear notching
1	Marking-permanent	Yes	Yes	No	Yes
2.	Number seen from distance	No	Yes	Yes	Yes
3.	Needs handling of animal for reading number	Yes	No	No	No
4	Reduces quality of hide	No	Yes	No	No
5.	Deforms the body part of marking	No	No	Partly	Yes
6.	Suitable on dark animals.	No	Yes	Yes	Yes
7.	Comparative cost	Least	less	less	less
8.	Precautions during procedure	Least	more	Less	more
9.	Care after milking	Least	more	Less	more

Other methods of identification

1) Use of leather neck strap
2) Keeling i.e. painting horns
3) Photographing
4) Muzzle prints

2. Dehorning

Removal of horn buds (in small calves) or horn (in adults) of animals is called as dehorning or disbudding.

Purpose/ Reasons for dehorning

- dehorned calves are easy to handle and proper feeding of animals.
- Dehorned calves require less space at the feed bunk and on trucks

- Less risk of injury with dehorned calves & less chance for horn cancer.
- Protection to animals against injury due to fight.
- It is essential for animals kept in loose housing.
- Uniform appearance.
- Less floor space required.
- It prevents body and udder injuries by long, pointed horns.
  - Calves should be dehorned at a young age 10 days to 2 months of age.
  - If possible do not dehorn during fly season.

Methods of Dehorning

- Chemical
  - Liquids
  - Caustic sticks
  - Paste
- Spoons
- Gouges
- Tubes
- Hot irons
- Barnes-Type
- Clippers
- Saws

1. Chemical dehorning (or disbudding)

Procedure

Secure the calf and throw gently on the bedding. Turn the head slightly towards operator. Locate the horn bud. Clip the hair 2 cm around the horn bud. Rub the horn bud with a piece of cotton wool soaked in surgical spirit. Apply Vaseline in a ring shape around the horn bud. Hold the caustic in the holder or with a piece of paper or cotton and wet the tip. Rub it briskly in a circular motion on the horn bud. Stop it as soon as the entire bud surface becomes reddish in appearance. Wipe the surface with cotton. Put some dusting powder. Treat the other horn bud similarly.

Optimum age of disbudding in calves = 2 weeks.

Substitute chemical for caustic potash.

1. Dehorning paste-used for destroying the horns matrix in calf hood. These often consist of Sodium hydroxide to prevent horn growth. Widely used paste formula contains 42% NaoH, 14% Ca $(OH)_2$, and 44% water.
2. Other chemicals- New patent formula Antomony tri-chloride. Salicylic dissolved on flexible collision.

Precautions

1. In case of bleeding, seal it with Tincture benzoin or Tincture ferriperchloride.
2. Take care and should not be turned out into rain after the treatment with caustic potash to prevent spreading and burning too large an area.
3. Caustic potash stick should be previously wrapped in paper to avoid burning of fingers.

2. Mechanical method of dehorning

It should be preferably being done in cold weather.

Procedure

Animal should be casted and thrown on the ground properly. When animals have partly or fully grown horns, the horn pincers or clippers or dehorning saw is used to cut the horn. The operation should takes place when animal is around 2 years of age or older. The wound should be covered with sulphanelamide powder mixed with iodoform or it may be treated with pinc tar or cotton soaked in pinc tar and then bandage.

3. Electrical method of disbudding

It is performed at three weeks of age in calves. It is quite safe and quick method. It is the most popular method.

Procedure

Secure the calf and thrown on the proper bedding gently. Locate the horn buds properly. Clip the hairs 2 cm around the buds properly. Switch on the current to make end of the electrical dehorner red hot (temp. 540 ^{0}C). The horn is cauterized by applying electrical dehorner just for 8 to 10 seconds. The calf is let loose when golden colour appears at the site of cauterized horn buds.

4. Rubber band method

It is used in calves having soft and small horns (5-10 cm).

Procedure

Secure the calf. Turn the calf's head slightly towards the operator. Make shallow grooves around the base of horn forming a ring. Slip a tight rubber ring over the horn with the help of elastrator and fix it into the groove. After few days the horn will gradually get out and fall on the ground because the tight rubber ring will shut off the blood supply on the horn.

It is not dependable and satisfactory method. It is painful method.

3. Castration

- Castration is the unsexing of the male or female or is the removal consists of both testicles and ovaries respectively. OR dysfunctioning of reproductive organs is called as castration
- **Benefits of castration-**
  1. Prevent unwanted reproduction.
  2. Leads to faster weight gain.
  3. Castration leads to produce more desirable type of meat.
  4. Make the animal docile and easier to handle.
  5. Castrated males can be housed along with heifers and cows.
  6. Veal/beef of castrated male is of superior quality.
  7. It is important in the treatment of orchitis, tumor of testis, spermatic cord, accidental injuries.
- **Best time for castration is-**

  For blood less castration- 4 to 6 months

  For incision method - 8 months
- **Methods of castration**
  - Knife
  - Burdizzo Castrator
  - Rubber/Elastor bands

Knife Castration

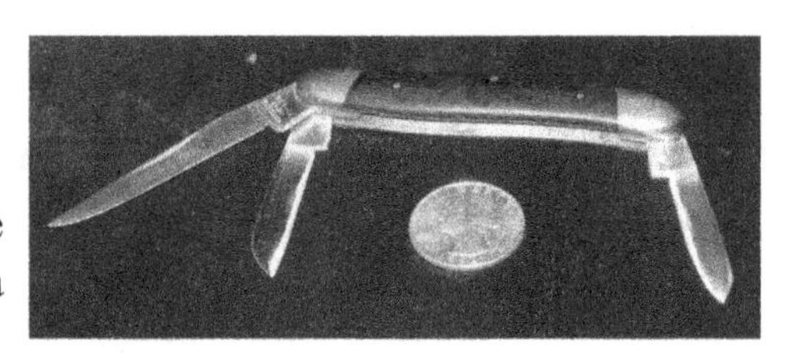

- Most widely used.
- Should only be done during a time of year when flies are not a problem.
- Calves should not be more than 3-4 months old.

Procedure

Secure the calf on clean floor and make it lie down on one side. Make hands and castrating knife clean and sterilize with spirit soaked cotton. Wash the scrotum in an antiseptic solution of acriflavin. Make an incision on the lower third outside of scrotum with sharp sterilized knife. Incision must extend well down to the end of scrotum and allow proper drainage of blood. Remove the testis by pulling them out, bring with them as much of the cord as possible. Scrotum and surrounding parts must again thoroughly disinfected with Tr. Iodine. Apply little Sulphanilamide powder mixed with iodoform.

- Results in an open wound
  - This increase the danger of infection and bleeding.
- Wound should be treated with iodine.
- Calves should be check several days after castration to check for swelling, continued bleeding and stiffness.

Burdizzo Castration

Procedure

Secure the calf and throw it on clean floor. Make the calf lie on one side and secure the legs. Manipulate testis and slightly pulled out with scrotum. Hold the spermatic cords tightly on both sides making sure that spermatic cord does not slip. Apply Tr. Iodine at the site of srush. Give local anaesthesia. Take the Burdizzo's castrator with clean and smooth edge of jaws, place it over the spermatic cords and punch quickly. The same process may again be repeated about 1 cm below first crush. Watch the animal for few days for any infection.

- Bloodless castration

- Crushes the cords of the testicles
  - However if the pincers are not applied correctly the cord may not be crushed completely resulting in a staggy steer later on.
- No open wound.
- Good choice in areas where screw worms are a problem.

Rubber /Elastrator Band Castration

- Special instrument that places a tight rubber band around the scrotum above the testicles.
- Cuts of the blood supply to the testicle.
  - This causes the testicle to waste away due to lack of blood
- No open wound.

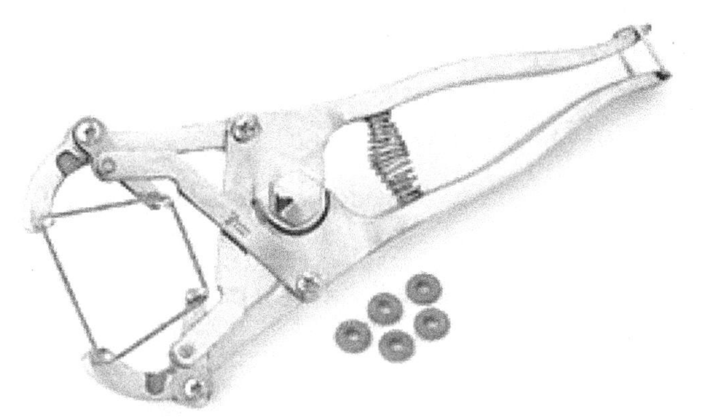

Rubber /Elastrator Band Castration

This method is used for sheep and goats at about 2-3 weeks of age. This is called as elastrator method.

Procedure

Secure the animal and makes it lie down one side on the clean floor. Select a tight rubber band. With the help of elastrator, place it over the spermatic cord of scrotum little above the testis on the scrotum. The constant pressure exerted by rubber shuts off the blood supply to testis. Testis will get dissolved and absorbed and rubber band slips and falls on the ground.

4. Grooming of animals

- Grooming is the brushing of body hair coat of animals. Also called brushing
- Equipments required for grooming
  1. Blunt type –comb
  2. Heavy brittle brush
  3. Coarse rope (made from paddy straw, coconut straw or dried grass

* Movements of brush should be along the direction of body hairs.

Grooming is done with the help of brush and curry comb. Grooming includes brushing followed by combing. Grooming is started at neck behind ear and finished towards hind quarter. Face is not brushed but wiped with a moist, clean cloth. Brush and comb are cleaned after 4-5 sweeps or strokes of grooming.

In milking animals grooming should performed 2 hrs. before milking to avoid contamination of milk with hairs, dust or dung particles.

Importance of grooming

1. It removes dirt & dust, loose hairs from the body.
2. Keep hide pliable.
3. Helps in clean and hygienic milk production.
4. Improves blood circulation.

5. Weighing of animals

Importance of weighing

1. It gives idea about weight gain or loss of body weight.
2. The feeding of animals is based on body weight of animals. It decides feeding strategy.
3. It helps in alteration of management of animals.
4. Puberty/Reproduction is depends upon body weight of animals.
5. Minimize treatment waste-Dose of animal is depends upon body weight of animal hence weighing of animals gives accurate idea of drug doses to be used.
6. Monitor reproductive performance-closely monitoring the weight gain of offspring can help make breeding decisions to maximize reproductive performance.

Two Methods of weighing of animals

1. Direct method- on platform balance
2. Indirect method-by body measurement

Formula used for calculation of body weight by measurement

1. Heart Girth x body length/300=body weight in pounds
2. (Heart girth)^2 x length/660= body weight in kg

6. Exercising of animals

- Exercise 2 hrs daily to animals in open paddock.
- Exercise keeps animal healthy.
- It increases intake and digestibility of feed.
- Increases yield & growth rate.
- Avoid leg problems.
- Keeps animal free from stress.

Care to be taken while exercising the animal

i) Sudden and forced exercise may result into impaired semen production in bulls.
ii) It is not easy to harness buffalo bull in bull exerciser.
iii) Strenuous exercise especially under inclement weather must be avoided.

CHAPTER 7

SYSTEMS OF HOUSING DAIRY ANIMALS AND MAINTENANCE OF HYGIENE AND SANITATION AT DAIRY FARM PREMISES

Objectives of housing

1. To protect animals from extreme environment condition.
2. To protect animals from wild life and theft.
3. To provide maximum comfort to animals.
4. To increase the efficiency of labour.
5. To provide clean comfortable shelter.
6. To provide efficient manage mental practices.
7. For clean milk production.

Selection of site/ location for a dairy house

1. **Topography & drainage:-** A dairy building should be at a higher elevation than the surrounding ground to offer a good slope for rainfall and drainage for the washes of the dairy to avoid stagnation within.
2. **Soil type:-** Fertile soil should be spared for cultivation. Foundation soil as far as possible should not be two dehydrated or desiccated. Such a soil is susceptible to considerable swelling during rainy season and exhibit numerous cracks and fissures.

3. **Exposure to sun & protection from wind:-** Building should be constructed in north-south direction(i.e. maximum exposure to the sun in the North and minimum exposure to the sun in South) and protection from prevailing strong wind currents whether hot or cold. Building should be placed so that direct sunlight can reach the platforms, gutters and mangers in cattle shed.
4. **Accessibility:** Easy accessibility to the building is always desirable.(Situation of cattle shed by the side of main road preferably at a distance of about 100m).
5. **Durability & attractiveness:-** It is always attractive when the buildings open up to a scenic view and add to the grandeur of the scenery. Durability of the structure is obiviously an important criteria in building a dairy.
6. **Water supply:-** Abundant supply of fresh, clean and soft water should be available at cheap rate. Water source in the form of well, river, dam should be nearby.
7. **Surroundings:-** Area infested with wild animals & doccits should be avoided. Narrow gates, high manger curbs, loose hingers, protruding nails, smooth finished floor in the areas where the cow move and other such hazards should be eliminated.
8. **Labour:-** Regular supply of skilled, reliable, cheaper and honest labour should be available.
9. **Marketing:-** Dairy buildings should only be in those areas from where the owner can sell his products profitably and regularly(It should be well connected to main road).
10. **Electricity:-** Adequate and continuous electricity supply should be available at the site.
11. It should be away from crowded area i.e. city but accessible to market.
12. **Facilities, labour & food:**
    - i) Feed storage should be located at hand near the centre of the cow barn.
    - ii) Milk house should be located almost at the centre of the barn.
    - iii) Centre cross-alley should be well designated with reference to feed storage, the stall area and the milk house.

Types/Systems of Housing

1) Loose housing system &
2) Conventional barn system

1. Loose housing system

"Loose housing may be defined as a system where animals are kept loose except at the time of treatment"

Loose housing system consist of 4 units as,

1) Paved area- 20 to 30 sq.ft. for each animal
2) Unpaved area- 80 to 100 sq.ft. for each animal
3) Feeding area- 30 inch for each animal
4) Milking room/milking parlor

The shed may be cemented or brick paved, but in any case it should be easy to clean a) the floor should be rough; so that animals will not slip. The drains in the shed should be shallow and preferably covered with removable tiles. B) The roof may be corrugated cement sheet, asbestos or brick and rafters. Cement concrete roofing is too expensive.

Inside the open in paved area it is always desirable to plant some good shady trees for excellent protection against direct cold winds in winter and to keep cool in summer.

At one end of paddock, a covered shed is provided for shelter.

Milking of animal is done in a separate milking barn. Feed and fodder is offered in a common manger. Water is provided in a common water tank. Total area is protected by a compound wall or fencing. Bedded area (about 60 sq. ft./ cow) is provided to give comfort to cows.

Advantages of loose housing system

1. This system is most economical.
2. Construction is cheap or inexpensive.
3. It is easier to detect animals in heat.
4. Animals are free to move, hence feel comfortable.
5. Labour requirement is less.
6. Expansion of housing without much modification is possible.
7. It is most suited for warm region.
8. Cows are milked in milking barn and hence helpful in clean milk production.
9. Animals get optimum exercise which is extremely important for better health and production.

Disadvantages

1. Chances of spread of contagious diseases are more.
2. Total floor space required is more than barn system of housing.
3. Animals in heat may disturb the other animals in the loose barn.
4. Display of animals in herd is not proper.
5. Not suitable in low temperature (cold) region.

Other provision:- The animal sheds should have proper facilities for milking barns, calf pens, calving pens and arrangement for store rooms etc.

2. Conventional Dairy barn

"In this system, animals are housed under a shed."

The following barns are generally needed for proper housing of different classes of dairy stock on the farm,

a) Cow houses or sheds
b) Calving box
c) Isolation box
d) Sheds for young stock
e) Bull or bullock sheds

a) Cow sheds

Cow shed can be arranged in a single row if the number of cows are small; say less than 10 or in a double row if the herd is a large one. i.e. single row- less than 10 cow.

Double row- Large herd

Ordinarily, not more than 80 to 100 cows should be placed in one building. This housing system is commonly used on organized dairy farms hence also called as Conventional dairy barn. In the double row housing, barn system of housing is mainly of two types:-

i) Tail to tail (face out)
ii) Head to head (face in)

i) Tail to tail system: In this system, animals are arranged in head out manner. There is common passage between two rows called as central alley.

Advantages of tail to tail system:

1. Cleaning of the shed is easier due to central alley.
2. Milking and its supervision is convenient due to central alley.
3. Cows in heat or having any reproductive problems can be detected immediately.
4. Chances for spread of diseases are minimum.
5. Animals are not disturbed by each other as all are facing out.
6. Use of milking machine is convenient and efficient.
7. Cows can always get more fresh air from outside.

Disadvantages of tail to tail system:

1. Feeding requires more time.
2. If not cleaned properly, it gives bad display to the visitor.

ii) **Head to head system:** In this system, animals are arranged in a head to head manner. There is a common feeding passage at the centre between the rows.

Advantages of head to head system:

1. Cows make a better showing for visitors when heads are together.
2. The cows feel easier to get into their stall.
3. Sun rays shine in the gutter where they are needed most or gutter dries quicker than face out system.
4. Feeding of cow is easier both rows can be fed without back tracking.
5. It is better for narrow barns.

Disadvantages of head to head system:

1. Milking and its supervision is not much convenient.
2. Detection of heat and Gyanaco-clinical problems is not easy.
3. While milking animals get disturbed by other animals facing to it.

Component/structure of farm building

1. **Floor:** It should be made from impervious material. It should be easy to clean and remain dry and non slippery. Cement concrete flooring with grooves or paved with bricks is preferable. A slope of 3 cm should be provided from manger towards the gutter. An over all floor space of 65 to 70 sq.ft. per adult cow should be satisfactory.
2. **Walls:** The inner side of wall should have a smooth hard finish of cement, which will not allow any lodgment of dust and moistures. Corners should be

round.(For plains, dwarf walls about 4 to 5 ft. in height and roofs supported by masonry work or iron pillars will be best or more suitable.

3. **Roof:** Roof of the barn may be of asbestos sheet or tiles. Corrugated iron sheets have the disadvantage of making extreme fluctuations i.e. Roofing iron sheets become hot in summer and cold in winter. However, iron sheets with aluminum painted tops to reflect sunrays and bottom provided with wooden insulated ceilings can also achieve the objective. A height of 8ft at the sides and 15 ft. at the ridge will be sufficient to give the necessary air space to the cows. Ventilation continuous ridge should be provided in the centre.

4. **Stall design:** The two main types of dairy barn stalls are

    i) Stanchion stall

    ii) Tie stall

    i) Stanchion stall: It is one of the standard dairy cow stalls. It is equipped with a stanchion for fastening a cow in a place.

    Animals are secured with the help of neck chain or stanchions at the manger therefore this method is also called as 'stanchion barns'.

    It is important to provide for the comfort of the cows and to the line them upto that most of the droppings and urine go to the gutter.

    ii) Tie stall: It requires a few inches longer and wider than the stanchion stall. It is designated to provide greater comfort to the cow. Due to the larger size, the chain tie gives the cow more freedom.

    Instead of the stanchion, there are two arches, one on each side of the neck of the cow. The correct space between arches is 10-12 inches. This prevents the cow from moving too far forward in the stall.

    In this type stall, the arches and all other stall parts are kept lower than the height of the cows.

    Large cows and those with large udders get along better in them because of freedom they enjoy. It is not desirable to have a tie chain in small stall.

5. **Manger**: Cement concrete continuous manger with removable partitions is the best from the point of view of durability and cleanliness.

    Height of manger:

    High front manger- 1 inch to 4 inch

    Low front manger- 6 inch to 9 inch

    The height at the back of the manger should be kept at 2 ft. 6 inch to 3 ft.

    An overall width of 2 ft. to 2.5 ft. is sufficient for a good manger.

6. **Alley/working passage:**

   Tail to tail system- 5-6 ft.

   Head to head system- 4-5 ft.

7. **Manure Gutter:**

   - 2 ft. width with 1 inch cross fall

8. **Doors:**

   Single row shed = 5 ft. wide & 7 ft. height

   Double row shed = Width is 8 to 9 ft.

b) Calving box

100 to 150 sq. ft. /cow with ample soft bedding. It is the separate area provided to cows or buffaloes those are due to parturition. The pregnant animals can be shifted in calving box 4-6 days before parturition. It provide comfort to pregnant animal and new born calf. Avoids injury to pregnant animals due to fighting of animals.

Calving box

c) Isolation box

It is the separate area provided to ill or sick animals or animals suffered with contagious disease or weak animals. In isolation box 150 sq.ft. area for each adult cow should be provided. It is useful for to avoid spread of contagious diseases, for proper growth of weak animals etc.

Isolation box

d) Sheds for youngstocks

Sheds for youngstocks

1. Calves below 3 months age
   Area required– 20-25 sq.ft./calf
2. Calves from 3-6 months
   Floor space required-25-30 sq.ft/calf
3. 6 to 12 months- Floor space required 30-40 sq.ft./calf
4. Above 1 year = Floor space required 40-50 sq.ft./calf

Difference between Loose housing & barn system

Sr. no.	Loose housing system	Barn system
1.	Animal kept in loose open paddock	Animals are tied at neck in the well protected shed
2.	Less expensive	More expensive
3.	Display of herd is not attractive	Display of herd is attractive
4.	A common feeding manger is provided	Manger is separate for each animal
5.	Disease control rather difficult	Disease control is better
6.	Animals are at more comfort	Animals are not at much comfort
7.	Used less at organized farms	Used at most of the organized farms
8.	Labour requirement is less	Labour requirement is more
9.	Heat detection is easier	Heat detection is difficult
10	Individual feeding as per requirement is not possible	Individual feeding as per requirement is possible

Difference between tail to tail system & head to head system

Sr. no.	Tail to system	Head to head system
1.	Animals are arranged in facing out position	Animals are arranged in face in position
2.	Feeding is difficult and requires more time	Feeding is easier and requires less time
3.	Detection of gynaecological problems and heat is easier	Detection of gynaecological problems and heat is not easier
4.	Cleaning of shed is easier and requires less time	Cleaning of shed needs comparatively more time
5.	Animals are displayed in fair way to visitors	Animals are displayed in better way to visitors
6.	Supervision of milking is easy	Supervision of milking is difficult
7.	Gutters are less exposed to sunlight	Gutters are more exposed to sunlight
8.	Less danger of spread of disease	More danger of spread of disease
9.	More space is required	Less space is required
10	Cost of construction is less	Cost of construction is more

Sanitation at Farm premises

It is defined as measures for preserving health especially removal of sewage.

I) Drainage: Three sections

1. The part inside & outside building situated above ground level.
2. Underground drain pipes & fittings.
3. Cesspool, septic tank, other means of disposal.

General principles to lay a drainage system

1. The drain pipes:- Non-absorbent material with air & water tight joints.
2. Pipes should be laid in straight lines.
3. All changes of directions or gradient should have open chamber for inspection.
4. Right angled junctions avoided.
5. Gradient to drains-self cleansing.
6. Inlets of foul drains should be trapped.
7. Pipe:- should be of proper dimension.
8. All the entrances to drains should be outside the animal building.

9. Provision for inspection of drainage.
10. Never pass a drain under the building.
11. Separate drainage system for rain water.
12. Sewage water carrying drain: - provided with suitable ventilation-escape of foul air.
13. Pipes: smooth internal surface.
14. Provision of gully trap- between drain & sewage.

II) Testing of drains:

- Testing should be done to check gas tight, water tight & water seals.
- **Following tests should be carried out periodically**

  i) Air test, ii) Smoke test, iii) Hydraulic test iv) Chemical test, v) coloured water test

III) Sewage disposal

- Two methods

  1) Conservatory method

  2) Water carriage system

IV) Storage of manure

V) Collection & storage of liquid

VI) Disposal of solid manure-methods

A) Physical methods- incineration and Burial

B) Chemical method:-Hellebore, BHC, Na-fluosillicate, Borax and Other

C) Biological methods:- Spreading/drying, Turning over the surface, Close packing and Biogas

VII) Compost making

VIII) Animal excreta as a factor in spread of disease.

IX) Manure as a breeding material for flies.

X) Fly borne diseases: C. pyogens, Anthrax, summer mastitis.

CHAPTER 8

CARE OF ANIMALS AT CALVING AND FEEDING AND MANAGEMENT OF CALVES

Care of cow during calving

- Separate pregnant animal from other and kept in a stall or calving pen. Before cow is transferred to calving pen, the pen should be disinfected using phenyl or Potassium permanganate solution.
- *Keep your cows in comfortable surroundings*

The calving period is a critical phase in the lactation circle.

- A successful start for cows and calves depends very much on the cows surroundings during this period.
- It is strongly recommended that cows are separated from the rest of the herd 2 – 4 days before calving, and that they have a special box with sufficient space and comfort.
- The area should be at least 8 – 10 m^2 per cow. The floor should be covered with a deep straw mattress for comfortable calving. Spread wheat bhusa, paddy straw, saw dust which is clean, dry, soft and free from dust, fungus.
- Single pens with measurements of 3 x 4 or 4 x 4 meters are suitable, but it is also possible to keep several cows in one bigger pen.
- Three cows share a calving pen of 30 m^2. There is a connection to a second pen which facilitates cleaning. In this case, the whole front part with feeding fence and mangers is made as a swinging gate which can be opened for manure removal.

- Of course it is also possible to clean the pens from the back.
- Keep close watch during calving.
- In handling advanced pregnant cows, care should be taken to prevent them from being injured by slipping on stable floors or by crowding through doorways or by mounting cows or bulls that are in heat.
- She should not be frightened or allowed to fight with other animals.
- Avoid rough handling as abusing, kicking etc. spoils temperament of cow which once formed is difficult to eradicate.
- She must not allow for walk to long distance. Provide limited amount of exercise to maintain better appetite and little more thrifty and to remove stiffness of limbs.
- Assist during parturition if required or if required take help from veterinarian.
- Pay special attention to first- calf heifers, because they are the most likely animals in the herd to experience calving difficulty.
- Supply lukewarm drinking water.
- Protection from cold winds and extreme temperature.
- After birth when placenta expelled, prevent cow from eating placenta.
- Proper disposal of placenta by deep burring in ground.
- Cleaning cows body with clean and warm water.
- Feeding a warm crude sugar mixed with bran mash.

Care of neonates immediate after birth

1. Remove the mucous from nose & mouth to avoid respiratory distress. If the calf does not start to breathe, artificial respiration should be used by alternative compressing & relaxing the chest walls with the hands after lying the calf on its side.
2. Ligate the umbilical cord at 2 cm from umbilicus with the nylon & cut with sterile scissor or blade. Apply Tr.Iodine to naval to avoid infection through umbilicus.
3. Keep the calf in front of cow/buffalo for licking which will develop the maternal instinct. If you want to wean the calf immediate after birth, do not allow the cow to lick the calf. Clean the body of calf with soft cloth.
4. Clean the udder & help to calf for suckling or remove the colostrums from udder & drench with utensils as shown in figure. Be sure that the calf should get first colostrum within 2 hours of birth.

The antibodies or immunoglobulin's present in colostrums protects the calf against diseases and has a laxative effect. The rate of colostrums feeding should be 10% of the calf's body weight.

5. Weaning means separation of calf from dam immediate after birth or after 3 days of birth or after 3-4 months of age.
6. The calf is best maintained in an individual pen or stall for the first few weeks. This allows more careful attention to individuals. After 8 weeks of age, calf may be handled in a group.
7. Take birth weight of calf immediately after birth and identify the calf by identification marking.
8. Dehorn the calf within 15 days of age.

Floor space requirement for calves

Age	Covered area	Open paddock
upto 3 months	1 m^2	1-1.5 m^2
Upto 6 month	1-2 m^2	2-2.5 m^2
From 6 month to 1 year	2 m^2	3.5-4 m^2

Generally calves are kept in narrow, poorly air-conditioned, or damp boxes. Also, they are often neither fed nor supervised with the necessary care. This results in poor calf health and considerable financial losses. So proper space should provide for calves along with proper feeding management. Separate housing should provide to ill and healthy calves.

FEEDING OF CALVES

Colostrum feeding to calves

"Colostrum is the first milk of a dam immediately after parturition". It is the first feed for newborn. Colostrum should be fed to calf within 2 hrs. after birth because during this period antibodies present in the colostrums are absorbed intact due to increased permeability of the intestinal wall. Later on this rate goes on decreasing. It should be given in 2 to 3 divided doses in a day. The interval between feeding should be maintained constant to avoid digestive complications.

Composition of colostrums

Sr. No.	Constituents	%
1	Total solids%	23.9
2	Fat%	6.7
3	Protein%	14.0
4	Antibodies%	6.0
5	Lactose%	2.7
6	Minerals%	1.11
7	Vit. A, ug/dl	295

Rate of colostrum feeding: It should be fed @ 10% of the calf's body weight. Excess of colostrum can be used to feed the other calves.

Colostrum should be fed to the calf for as long as it is available, usually 3-4 days. If the cow died during the calving, then the colostrum from the another cow is may be used.

Importance of colostrum feeding

1. It contains antibodies named gamma globulins which provide immunity against calf hood diseases.
2. Its laxative action helps to remove meconium i.e. first faeces of calf.
3. It also creates acidic medium in digestive system and thereby prevents diarrhoea.
4. Colostrum contains about 3 to 5 times more protein than normal milk (about 20%)
5. Colostrum contains 5 to 15 times more vitamin A than that of normal milk.
6. It is also rich in riboflavin, choline, thiamine and pantothenic acid
7. It also supplies some important minerals like copper, iron, magnesium and manganese.

Colostrum substitute

1. If colostrum is not available due to death of dam or non-secretion from the udder, then colostrums from other freshly calved cow can be given.
2. A mixture of the following ingredients can acts as colostrums substitute in emergency-

   Water = 275 ml, Milk = 525 ml, Castor oil = 2.5 ml, Egg = one and auromycin powder 10-25 gm.

   In addition to this, injection of 50 ml of dam's serum intravenously is also recommended.

Feeding milk to calves

1. The milk is the food, which is ideally suited to digestive system of the calf.
2. The milk is called as a whole feed for a developing calf.
3. The milk should be fed at the rate of 10% of the body weight in first month, 15% of the body weight in second month and 20% in third month.

4. Milk should be fed either pale or nipple feeding or suckling method.
5. Milk feeding pail should be regularly washed and cleaned so as to avoid growth of micro-organisms.

Teaching of calf to drink

For teaching to drink patience will be required as the some calves are slow in learning to take milk from pail. One should pour about a part of the mother's milk into clean pail used for feeding calves and bring the nose of the calf in contact with milk. This is best accomplished by allowing the calf to suck the finger of the feeder so that its head may be guided into the pail and then the hand of the feeder can be gradually lowered into the bucket and submerged in the milk sufficiently deep to allow a little milk to be taken by the calf. By continuous feeding it will learn to drink.

In some farms it is practice to use pail with nipple to feed the milk to young ones. The calf sucks the nipple and milk enters in mouth of calf. The nipple pail has the advantage in that the calf takes the milk more slowly, and is thus less likely to have digestive upsets. Nipple pails should rinse thoroughly after each feeding.

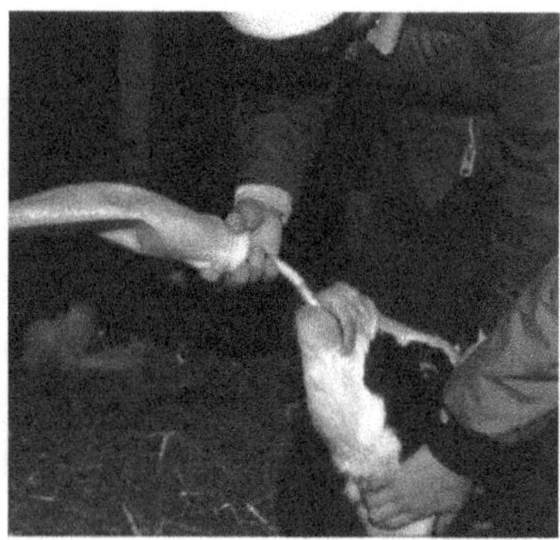

Calf to drink

Feeding of milk to calves

Milk is the whole feed for calf hence it should be necessary to feed the whole milk @ 10% of the body weight in first 21 days, @ 15% of body weight from 22nd to 35 day & @ 20% of body weight from day 36 to 2 month of age.

Feeding milk replacer to calves- Milk replacer is available in powder form.

- Milk replacer is usually fed in gruel form (1 part milk replacer + 4 parts of water)

- Benefits of feeding milk replacer
  1. Improves growth rate of calves due to high protein content.
  2. Reduces illness by improving health and immunity of calves.
  3. Reduces cost of calf rearing by making milk available for sale.
  4. It saves milk for utilization for human consumption.
  5. Reduces mortality of calves.

Chemical composition of Milk replacer

Sr. No.	Nutrients	Percentage
1.	CP (% min.)	20 to 28
2.	Ether extract/Fat (% min.)	10 to 22
3.	CF (% Max.)	1 to 2
4.	Minerals (%)Ca, P, Mg	1.00, 0.7, 0.07
5.	Vitamins(IU/lbs) A, D, E	4.091, 2.73, 22.7

Example-Milk Replacer Formula

Ingredients	Quantity (Kg)
Wheat	10
Fish meal	12
Linseed meal	40
Milk	13
Coconut oil	07
Linseed oil/Cotton seed oil	03
Citric acid	1.5
Molasses	10
Mineral mixture	3
Butyric acid	0.3
Antibiotic mixture	0.3
Rovimix-A, B2, D3	0.015

Feeding of Calf starter to calves

- It is the first solid feed given to calves.
- It contains CP 18 to 20%, Fat 3% & TDN 75%.
- It may fed from 2 weeks of age upto 4 month of age.

Advantages of Calf starter Feeding

1. Produces heavier calves at weaning (30-70 lbs).
2. Helps in earlier rumen development.
3. Increase the number of essential ruminal microbes which are essential for better digestion.

Example- Calf Starter Formula

Ingredient	Quantity (%)
Barley	40
Groundnut cake	25
Soybean cake	25
Dried skim milk	8
Steamed bone meal	1
Vitamins and minerals	1 (yeast –25g/100kg)

Feeding schedule for Calves (0-6 months age group)

Age (Days)	Colostrum (Lit.)	Milk (Lit.)	Calf Starter	Green Fodder
0-4	2-2.5 (1/10th of B.W)	–	–	–
4-30	–	2.5-3.0 (1/10th of B.W)	–	*Ad lib* preferably leguminous fodder (after 15 days)
30-60	–	3.0-4.0	50-100 gm	*Ad lib*
60-90	–	3.0-3.5 Fat separated Milk	100 gm to 250 gm	*Ad lib*
90-180	–	1.5-0.5	250-750gm	*Ad lib* (leguminous or non leguminous)

Fresh water for calves

1. Fresh clean water at all times, particularly when milk feeding is reduced or discontinued.
2. Watering to calves just before feeding milk should be avoided as it would reduce intake of milk.
3. There is positive relationship between water consumption and body weight of calves.

4. Calf is not satisfied with milk alone as a drink and it wants to drink quite often little water during day time.
5. Don't allow calf to drink to more water at one time it may leads to red urine.

Weaning of calves

Separation of calf from dam and fed artificially is called as weaning. Weaning can be practiced by removing the calf from dam immediate after birth or calf remains with dam for 2-3 months and after that removed to calf pen.

Advantages of weaning system

1. Cow continuous to give milk whether/ not its calf is alive.
2. Calf can be culled at an early stage.
3. With the help of milk substitute calves can be raised economically even if the dam dies.
4. Avoids risks of complications due to under feeding and over feeding a calf.
5. Exact amount of milk produced by cow can be determined.
6. Hygienic milk production is possible.
7. Cows become regular breeder.
8. Calving interval in weaned calves reported to be 13 to 14 months against 16 to 18 months in unweaned calves.

Difference between suckling and weaning system of calf rearing

Sr.No.	Suckling	Weaning
1	Means calf directly allow to suckle the dam	Means Separation of calf from dam and fed artificially
2	Continuous to give milk is not possible if calf is died because of development of strong maternal instinct	Cow continuous to give milk whether/ not its calf is alive
3	It is not possible that Calf can be culled at an early stage	It is possible that Calf can be culled at an early stage
4	It is not economical method of calf rearing	It is economical method of calf rearing
5	Labour requirement is less due to suckling	Labour requirement is more due to manual or artificially feeding to calves.
6	More chances of under or overfeeding of milk to calves	Avoids risks of complications due to under feeding and over feeding a calf

[Table Contd.

Contd. Table]

Sr.No.	Suckling	Weaning
7	It is not possible to record exact quantity of milk production of cow due to suckling	The exact quantity of milk production by cow can be recorded easily.
8	Hygienic milk production is not possible	Hygienic milk production is possible
9	Cows may become repeat breeder or not come in estrus for long duration	Cows become regular breeder
10	There may be chances of teat wounds due to suckling by calf	There is no chance of teat wound

Health care of calves

1. **Deworming of calves:** After birth the calf exposes to environment, dirt, dust dung of other animals and worm infestation occurs. The endoparasites utilizes digested feed in the body of host, also some parasites sucks the blood and leads to stunted growth of calves or mortality of calves. Hence, first deworming should carried out on 8 to 10 days of birth & then after every month upto 6 months of age. After 6 month of age **deworming** should be carried out after every 3 months routinely after faecal sample examination. Example of dewormers- Albendazole, Fenbendazole, Piperazine, closentel etc. Use available anthelmintics alternatively for deworming of animals.
2. Control of external parasites by spraying, applying organophosphate insecticides (External parasites sucks the blood leads to anemia, also tickborne protozooan diseases).
3. **Vaccination in calves**

Sr.No.	Name of Disease	Type of vaccine	Time of vaccination	Duration of immunity
1.	Anthrax	Spore vaccine	Pre-monsoon	One season
2.	BQ	Formal killed vaccine	Pre-monsoon	One season (6 month)
3.	HS	Oil Adjuvant vaccine	Pre-monsoon	One season (6 month)
4.	Brucellosis	Cotton strain 19	About 6 month of age	3 to 4 calvings
5.	FMD	Polyvalent tissue culture vaccine	At 4 month of age	After 1st vaccination repeat vaccination every year in oct. or Nov.

Repeat the vaccine depending upon duration of immunity for different vaccines manufactured by different companies.

4. **Removal of extra teats:** Extra teats more than 4 are removed when the calf is between 1 to 2 months of old. Extra teat should be clipped off with a pair of sterilized scissors and a disinfectant such as Tincture iodine must be applied.

CHAPTER 9

MANAGEMENT OF LACTATING AND DRY COWS AND BUFFALOES

Management

Immediate after parturition wash the exterior genitalia, flanks and tail with warm water or water containing crystal of potassium permagnate or neem leaves boiled in water.

Keep the cow warm- provide warm water to drink, warm gur sarbat to drink just after parturition (specially in winter).

- Dispose off the placenta by deep burial.
- Avoid ingestion or licking of placenta by animal.
- Remove the placenta after 12hrs of parturition.
- Gaurding the animal from extreme of temparature in summer.
- To avoid milk fever, do not draw the all milk from udder for a day or two days after calving.
- Regular checkup for mastitis.
- After parturition when cow is milked first time, milker must ensure that all blockages from teat are removed.
- For first few days, milking should be done for three times a day to avoid udder congestion or edema.
- Regular cleaning of shed.
- Routine faecal sample examination & deworming.
- Routine vaccination of cattle & buffaloes against FMD, HS & BQ.
- Avoid to much crowd in shed.

- Keep animal in well ventillated sheds.
- If the udder is somewhat hard and inflamed there is no need to be alaramed so long milk can be obtained from each quarter freely.

Feeding of Lactating cows & buffaloes

- Immediately after calving, mild luxative, palatable and energy giving feeds should be given.
- 2 kg bran + 1 kg jaggery or mollasses moistened with water.
- A little quantity of palatable and easily digestible green fodder should be given
- Provide mineral mixture to avoid milk fever in high yielding animals.
- Provide good quality green fodder like Maize, Perrenial grasses, leguminous fodders etc.
- During early lactation, appetite of animal is reduced. The animal is in negative energy balance, looses weight.
- So to provide all essential nutrients the diet should formulate with high energy, or by increasing the nutrient density of feed.
- Provide 50-100 gms of mineral mixture depending upon milk yield of cow/ buffalo and also provide 40 gms of common salt.
- Provide Bypass protein, Bypass fat to high yielding animals during peak production.
- A fresh cow should be fed increased amount of feed to lead to full production untill she no longer responds to additional feed and again reduce the feed to level of maintaining peak yield.
- Guard the animal against milk fever & Ketosis.

Concentrate feeding to milking cows & buffaloes

- 1 kg concentrate mixture for 2.5 kg milk over maintenance requirement to zebu cattle and per 2kg milk in case of buffaloes.
- Concentrate mixture should contains 20% CP and 65 % TDN.
- Concentrate mixture should contain required quantity of important minerals like calcium and phosphorus.
- Pelleted feed is generally preferred than mash feed for milking animals.
- At about 15 days of calving, give slightly more quantity of concentrate than actual production which is called as Challenge feeding or lead feeding.

- A concentrate mixture should be sprinkled with water to lessen its dustiness before feeding at the time of milking.

BIS specifications for cattle feed for milch animals

Nutrients	Type I	Type II
Moisture, % Max	11	11
Crude Protein, % min.	22	20
Ether extract, % min.	3	2.5
Crude fibre, % max	7	12
Acid insoluble ash, % max	3	4

MANAGEMENT OF DRY COWS

Need of dry period

1. To rest the organs of milk secretion.
2. To permit the nutrients in the feed to be used in developing the foetus instead of producing milk.
3. To enable the cow to replenish in her body the stores of minerals which have become depleted through milk production.
4. To permit her to build up a reserve of body flesh before calving
   - Always separate the dry cows from the milking herd.

The effect of the length of the dry period on production is influenced by the body condition at drying off and by feeding practices during the dry period.

- Feeding practices during the dry period can greatly affect production in the next lactation.
- The dry period should be of 60-90 days.
- Drying off- stop the lactation of cow or buffalo 60-90 days before calving.
- Methods of drying off
  1. Incomplete milking
  2. Intermittent milking
  3. Complete cessation
- Routinely observe dry cows for problems and watch closely at calving. Provide a clean, dry place for calving.

Feeding of dry cows

- Dry cows require smaller amounts of nutrients than lactating cows.
- The feeding program depends on the condition of the cow at dry-off time.
- Do not over-condition before or after the cow turns dry.
- Cows that are too thin will need extra feed to prevent limiting subsequent production.
- During dry period provide minerals and vitamins through mineral- vitamin mixture.

Feeding during the late dry period

- Higher grain level should be fed before calving so the rumen microflora will be adjusted to the high grain ration the cow will receive from calving.
- During the final 10 days before calving; increase the grain so the cow is eating Vz to 500-1000 gm per 100 kg of body weight by calving. This is called steaming up.
- To reduce milk fever, the grain mixture should have less calcium than in the normal lactating cow mixture.
- At 3 to 4 weeks before calving, run first-calf heifers through the parlor several times with the regular milking herd to adjust them to that environment.

CHAPTER 10

IDENTIFICATION COMMON FEEDS AND FODDER OR CLASSIFICATION OF FEEDSTUFFS

Objectives

1] To know the various feeding resources for dairy animals.

FEED STUFF

A] **Roughages-** These are the feedstuffs containing more than 18% Crude fiber & less than 60% TDN.

1] **Succulent Roughages-** These are the feedstuffs with moisture 60 to 90%.

i] **Pasture**

ii] **Tree leaves** [Pipal, Bel, Katchnar, etc]

iii] **Silage**

iv] **Root crops** [Carrot, Turnio, Tapioca, Fodder Beet etc.

V] **Green fodder-**

a] **Leguminous green fodder**- [Cow pea, Cluster bean, Green pea, Berseem, Lucerne, etc]

b] **Non-leguminous green fodder**-[These are the cereal fodders. Fodder of Jowar, Maize, Bjara, Oat, Barley, Sudan grass, Napier grass, Para grass, Pauna, DHN6, NB21 etc]

2] **Dry Roughages-**

i] **Kadbi /Straw-** [Jowar kadbi/straw, Rice straw, Wheat straw, Oat straw, Cow pea straw, Groundnut straw, Gram straw, Mung straw, etc]

ii] **Hay**

a] **Leguminous** hay- [Lucerne hay]

b] **Non-leguminous** hay- [Hay of Sorghum]

B] **Concentrates-**

1] **Energy Rich-**

i] **Grains and seeds-** [Maize, Barley, Sorghum, Oat]

ii] **Mill byproducts** [Rice bran, wheat bran, etc]

iii] **Roots-** [Cassava, potatoes, sweet potatoes, etc]

iv] **Molasses-** [Sugar cane molasses, Wood molasses, Citrus molasses, Beet molasses]

2] **Protein rich**

i] **Animal origin-** [Meat meal, blood meat, meat scrap, etc]

ii] **Marine byproducts-** [Fish meal]

iii] **Avian byproducts-** [Feather meal, hatchery by-products meal]

iv] **Plant origin**: Oil **seed meal, Oil cake-** [Ground nut oil meal, Linseed meal, Soybean oil meal, etc., Cotton seed cake, Groundnut cake etc.]

v] **Non protein nitrogenous substances-** [Urea, biuret, oil seed meals, Ammonium acetate, Ammonium bicarbonate, Ammonium lactates]

vi] **Brewers grains and yeast-**

vii] **Single cell protein** [Bacteria like Methanomonas methanica, algae like Chlorella vulgaris and Scenedesmus obliquus, yeast like Saccharomyces cerevisiae, Torulopsis utilis, and Candida lipolytica]

C] **Feed Supplements** – 1.Mineral suppliments=[Various minerals Ca, P, Fe, I, Mg, Cu, etc]

2. **Vitamin suppliment-** [A, D, E, K, B. complex, C, etc]

D] **Feed Additives-** [Antibiotics, Growth promoters, Probiotics, Prebiotics, hormones, colouring agents, flavouring agents, medicants]

Sr.No.	Roughages	Sr.No.	Concentrates
1	These are the feedstuffs contains more than 18% crude fibre & TDN value is less than 60%.	1	These are the feedstuff contains less than 18% crude fibre & TDN value is more than 60%.

[Table Contd.

Contd. Table]

Sr.No.	Roughages	Sr.No.	Concentrates
2	Roughages are bulky generally.	2	Concentrates are less bulky.
3	Density of nutrient is less per unit volume	3	Density of Nutrient is more per unit volume
4	Roughages have low weight per unit volume.	4	Concentrates have more weight per unit volume.
5	Roughages are classified as dry & succulent	5	Concentrates are classified as Energy rich concentrates & protein rich concentrates.
6	They are cheaper in cost.	6	Concentrates are costly than roughages.
7	Digestibility is less than concentrates.	7	Digestibility is more as compared to roughages.
8	Less nutritive value than concentrates.	8	More nutritive value than roughages.

	Feed additive	Feed supplements
1	These are the substances or chemicals added In animal feed for the purpose of improving. Rate of gain, milk production, feed efficiency, prevention of disease & increasing storage Value & acceptability of feed e.g. Antibiotics Probiotics, antioxidants.	These are the substances which supplement the action of roughages/ concentrates e.g. mineral & vitamin supplements.
2	Feed additives are generally non-nutritive.	Feed supplements are nutritive value
3	Feed additives are not dietary essential.	Feed supplements are dietary essential
4	Not use in daily feeding schedule.	It is essential to provide feed supplements daily.
5	Feed additives are used during emergency with specific objectives	Feed supplements are used regularly to satisfy the nutrient requirement of animal for mineral & vitamin.
6	Feed additives are used in minute quantity in Feed.	Feed supplements are used in more quantity than feed additives.

	Legume fodder	Cereal fodder
1	Legume fodders are rich in protein	Cereal fodders are rich in energy
2	Legume fodders contain more calcium as compared to cereal fodder	Cereal fodders contain less calcium as compared to legume fodder
3	Most of the legumes are of thin stem	Most of the cereals are of thick stem
4	Palatability is more	Palatability is less as compared to legume fodder
5	Legume fodder is fed in less quantity as compared to cereals	Cereal fodder is fed in more quantity as compared to legumes.
6	Legume fodder is of less height	Height of cereal fodder is more than legumes
7	These are less suitable for silage making, used in composite silage	These are more suitable for silage making
8	Legume fodder fixes environmental nitrogen and increases nitrogen content of soil	Cereal fodders can not fix the environmental nitrogen
9	More digestible than cereals	Less digestible than legumes
10	High feeding may leads to bloat	Less chances of bloat
11	e.g. Lucerne, cowpea, Berseem etc.	e.g. Maize, Jowar, Bajra etc.

CHAPTER 11

PREPARATION OF RATIONS FOR ADULT ANIMALS

In preparation of ration for adult animals, important matter is to ascertain and to meet up total requirement of dry matter, digestible protein [DCP] and total digestible nutrients [Energy] i. e. TDN for period of 24 hours.

Total dry matter

Requirement of dry matter depends upon body weight of animal and physiological status. Generally Zebu cattle need 2 to 2.5 kgs, dry matter for each 100 kgs. Live weight. In buffaloes and crossbred animals it is 2.5 to 3 kgs. /100 kgs. Body weight. This total dry matter should meet the requirement of carbohydrate, protein, fat, minerals, vitamins. This dry matter allowance should be offered in following way-

Total dry matter = 2/3rd DM from roughages & 1/3rd DM from concentrate

[Out of which **2/3rd DM dry roughage** or ¾th DM through if legume is available sufficiently & **1/3rd DM through green roughages** or 1/4th if leguminous fodder].

Maintenance ration

This is the minimum amount of feed required to maintain essential body processes at their optimum rate without gain or loss in body weight or change in body composition. It should provide 14 to 16% digestible crude protein [DCP] and 68 to 72% TDN. For this, along with roughages, zebu cow should get 1 to 1.25 kg. Concentrate while crossbred cows, purebred Indian cow and buffaloes should be given 2kg. Concentrate mixture.

Gestation ration

Additional allowance over and above the maintenance ration should be given to the cow after seventh month of pregnancy for growth of foetus and to keep the cow healthy for future production. Decide an amount of concentrate feeding depending on size of cow, condition of cow and roughage quality. In general, in addition to maintenance ration, additional amount of 1.25-1.5 kg. Concentrate mixture is recommended for zebu cows and 1.75-2.5 kg for crossbred cows, buffaloes. Cow besides her maintenance and milk production requirement should get additional amount of 0.14 kg, DCP and 0.67 kg. TDN along with essential minerals and vitamins. High yielding cows/ buffaloes should have liberal access for feeding after eight month of pregnancy for ensuring development of mammary gland.

During gestation animal should be liberally fed and treat kindly. The advance pregnant animal should provide slightly luxative feed with sufficient minerals and vitamins and with sufficient protein. Animal should provide preferabally leguminous fodders. Sole feeding of dry fodder should be avoided.

Steaming up

- The 60-70 percent growth of calf in womb is occurred in last 2 to 3 months. Hence feeding during this period is important.
- Feeding a dry cow specially in preparation for calving is termed as "steaming up"

  For this purpose cow must be fed 0.15 kg DCP and 0.45 kg TDN.

Advantages of steaming up

1. Increases daily milk yield after parturition or calving.
2. Slightly lenghten the period of lactation.
3. Promotes growth of mammary tissue.
4. Slightly increases butter fat percentage.
5. Increase the body reserve.

   Make available clean drinking water for 24 hrs.

Production ration

Additional allowance of ration for milk production over and above maintenance requirement is given to the lactating animal. As a thumb rule, for Zebu cow for every 2.5 kg milk, one kg additional amount of concentrate is required over and above maintenance requirement. In case of buffaloes, purebred cows and crossbred cows this propotion should be one kg concentrate over and above maintenance need for every two kg milk production. A cow giving milk having 4% fat needs 0.045 kg. DCP and 0.316 kg TDN per kg. of milk produced.

Requirement for breeding bull

To maintain reproductive health of bull in good condition, one kg additional concentrate mixture should be given over and above maintenance need. The requirement varies with breed, size, age. An adult fully grown bull [Approximate weight 500kgs.] needs 0.45 kg. DCP. 4.05 kg. TDN with calcium and phosphorous 11 grams each per day. Ration for bull should be able to provide 12% DCP and 70% TDN.

Feeding of growing bull: - The ration of young growing bull must contain DCP 12-15% and TDN 70%. Legume hay may be given @ 1 kg per 100 kg body weight. Also provide concentrate 1.5 to 2 kg depending on breed and size. Amount of roughages will depend upon its quality and size of a bull. Young bulls receive sufficient greens to meet the needs of vitamin A; deficiency of which may lead to poor quality semen production. Ration must contain sufficient minerals. If young bull fed higher levels of TDN (115-130%) develops weakness of feet and legs at 3 years of age.

Feeding of bull in service

Bull in service should be neither fat nor thin. Ration containing sufficient protein (10 to 12 % DCP and 70% TDN), minerals and vitamins. If the roughages include legume then concentrate containing 10% DCP may be given. Bulls are not usually pastured because of difficulty in controlling them. Dry matter may be given @ 2 kg/ 100 kg body weight. During heavy breeding season the concentrate allowance should be increases and vice-versa. While feeding bull, care should be taken to avoid phosphorus deficiency which leads to sterility problems. Excess of calcium in bulls diet may cause vertebrae and other bones to fuse together leading to poor mounting, hence may be avoided.

Requirement of bullock-

Bullock weighing 500 kg Body weight needs 0.56 kg. DCP and 4.5 kg TDN for normal work and 0.71 kg. DCP and 6.4 kg. TDN for heavy work. In addition, bullock should be offered salt and mineral mixture because during drafting operation due to muscular contraction, sodium chloride is lost through perspiration and calcium, phosphorous and magnesium through intestine and kidney.

Some ration formulae

1] For milking cows/buffaloes-

Ingredient	Proportion[%]
Cotton seed cake	30
Maize chuni	55
Udid chuni	12
Mineral mixture	02
Salt	01
Total	**100%**

ii] For pregnant cows/ buffaloes-

Ingredient	Proportion[%]
Cotton seed cake	52
Maize chuni	30
Udid chuni	15
Mineral mixture	02
Salt	01
Total	**100%**

Steps for preparation of ration:

1. Calculation of Dry matter requirement of animal for maintenance or maintenance+ production or maintenance + pregnancy.
2. Partitioning of total dry matter into roughages & concentrates.
3. Calculate the nutrient requirement of animal (DCP & TDN).
4. Determine the amounts of available ingredients that must be fed to fulfill the animals nutrient requirements within its expected dry matter intake limit.
5. Supply mineral mixture & salt @ 1% of total quantity of concentrate required.

Balanced ration

Balanced ration is the allowance of feeds & fodders given to the animal for 24 hours & which provides the essential nutrients to the animal in required quality, quantity & proportion.

DESIRABLE CHARACTERISTICS OF A BALANCED RATION

Characteristics of balanced ration

1. **Liberal feeding:** A liberal amount of feed should be given to the animal so as to satisfy the nutritive requirement. However excessive feeding should be avoided as it leads to wastage and economic loss.
2. **Individual feeding:** As nutrient requirement differs from animal to animal, it is necessary to feed the animal individually e.g. during late pregnancy or early lactation nutrient requirement increases which can be satisfies by extra feeding. So that the foetal growth and lactation yield is maintained respectively. Similarly where number of animals are more, groups can be made depending upon the physiological status and age of the animal and then group feeding can be performed.
3. **Ration should be slightly laxative:** It is necessary that ration should be slightly laxative as it maintains the normal passage of ingesta and keeps the digestion normal. However if ration is highly laxative digestibility of nutrient is reduced and problems like diarrhoea and dysentery may start. If ration is low laxative then digestive problems like constipation develops.
4. **Ration should be palatable:** Palatability of ration has direct influence on the voluntary feed intake, therefore feed with low palatability should not be considered for ration preparation inspite of its higher nutritive value. The palatability of ration can be improved by addition of molasses, jaggery or salt.
5. **Ration should be free from moulds, fungus and toxic substances:** As the feedstuff with mould, fungus and toxic substances affects the nutrient availability, health and production of animal. Such feedstuff should be avoided or necessary processing should be done to make these feed ingredients safe for animal consumption before they are included in balanced ration.
6. **Ration should include variety of feedstuff:** The combination of different feedstuff makes better availability of nutrients. The balanced ration should therefore include variety of cereals, cakes, brans for preparation of concentrate mixture and green fodder. Cereal fodders, leguminous fodders should be used to constitute roughage portion of the diet.
7. **Ration should contain enough amount of mineral matter:** Minerals play an important role in the body. Therefore its supply through ration should be assured, otherwise deficiency results. This can be achieved by including feedstuff rich in mineral content or by use of mineral mixture and common salt in the concentrate mixture.

8. **Sudden change in diet should be avoided:** Any sudden change in diet may produce digestive complications like diarrhoea, indigestion, constipation or tympany. Therefore any change in the diet of animal should be made gradually.
9. **Maintain regularity in feeding time:** The regular feeding time keeps the constant feeding interval and thereby avoids digestive complications. The regular feeding time assures sufficient glandular secretions like saliva and fixed feeding time that leads feeling hunger followed by better digestion.
10. **Ration should be properly processed:** Proper processing of ration ingredients is necessary to improve its acceptability and utilization.

    e.g. cereal grains should be coarsely grinded and cereal fodder should be chaffed.
11. **Ration should be fairly bulky:** A fairly bulky ration is essential as it gives satisfaction of hunger and therefore while preparing a balanced ration care should be taken to provide fair amount of bulk in addition to essential nutrients.
12. **Include much of green fodder in the diet of animal:** The green fodder gives many advantages to the animal. A green fodder gives cooling effect to the digestive tract. It is good source of carotene. It is palatable, digestible contains unidentified growth factors and also lengthens the productive life of the animal.
13. **Ration should be least cost:** While formulating a balanced ration cost involved on purchase of feed ingredients and processing should be taken into account so that the cost of ration is minimized.

Computation of ration

Example. 1

Compute a ration for a cow weighing 600 kg and producing 25 litres of milk per day with 4% fat

Available feed & fodder with composition

	DM	DCP	TDN
DHN 6	25%	10%	60%
Kadbi	90%	0%	42%
Concentrates	90%	18%	75%

Weight of cow = 600kg

Thus, DM required = 3% of body weight

i.e, DM required = 18%

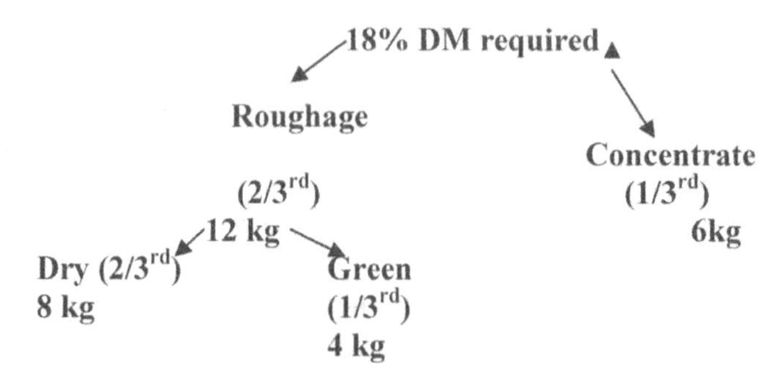

Calculation of DCP and TDN requirement of animal

	DCP	TDN
Maintenance	0.350kg	3.804kg
Production	1.125kg	7.904kg
Total	1.475 kg	11.70 kg

Calculation of DCP and TDN to be supplied:

Feed/fodder	DM to be supplied	DCP to be supplied	TDN to be supplied	Actual quantity of Feed/fodder to be supplied
DHN 6	4 kg	0.4 kg	2.4 kg	16 kg
Kadbi	8 kg	0 kg	3.36 kg	8.88 kg
Concentrate	6 kg	1.08 kg	4.5 kg	8.33 kg
Total		1.48 kg	10.26 kg	

Salt and Mineral mixture to be supplied

Salt and mineral mixture to be supplied through ration=1% of the total quantity of concentrate mixture

Thus, salt and mineral mixture should be fed = 83 gm each/day

Example. 2

Compute a ration for a buffalo weighing 300 kg and producing 6 litres of milk per day with 7% fat.

Feed & fodder Available for feeding & their composition

	DM	DCP	TDN
DHN 6	25%	10%	60%
Kadbi	90%	0%	42%
Concentrates	90%	18%	75%

Weight of cow = 300kg

Thus, DM required = 3% of body weight

i.e, DM required = 9%

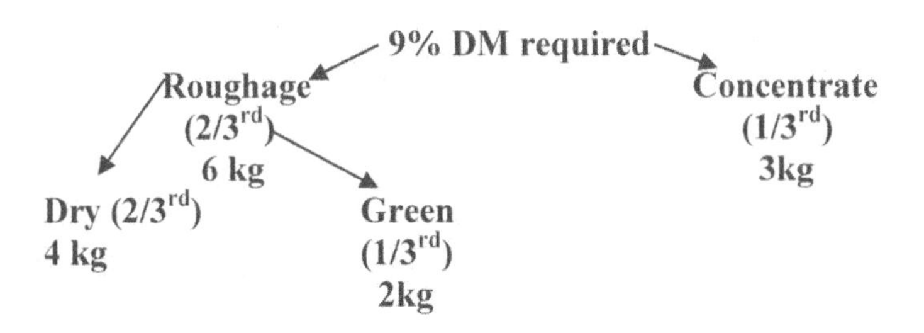

Calculation of DCP and TDN requirement of animal

	DCP	TDN (for 6 ltr. milk)
Maintenance	0.197kg	2.36kg
Production	0.378kg	2.748kg
Total	0.575 kg	5.108 kg

Calculation of DCP and TDN to be supplied:

Feed/fodder	DM to be supplied	DCP to be supplied	TDN to be supplied	Actual quantity of Feed/fodder to be supplied
DHN 6	2 kg	0.2 kg	1.2 kg	8 kg
Kadbi	4 kg	0 kg	1.68 kg	4.44 kg
Concentrate	3 kg	0.54 kg	2.25 kg	3.33 kg
Total		0.74 kg	5.13 kg	

Salt and Mineral mixture to be supplied

Salt and mineral mixture to be supplied through ration=1% of the total quantity of concentrate mixture

Thus, salt and mineral mixture should be fed = 33 gm each/day

Example 3

Compute a ration for an indigenous cow weighing 400 kg and producing milk 10 liter with 4 % fat

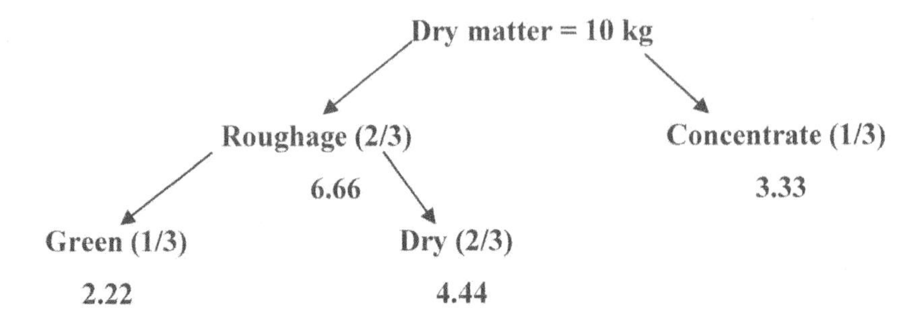

Available feed & fodder with composition

	DM	DCP	TDN
DHN 6	25%	10%	60%
Kadbi	90%	0%	42%
Concentrates	90%	18%	75%

Calculation of DCP and TDN requirement of animal

	DCP	TDN (for 10 ltr. milk)
Maintenance	0.254 kg	3.03 kg
Production	0.45 kg	3.16 kg
Total	0.704 kg	6.19 kg

Calculation of DCP and TDN to be supplied:

Feed/fodder	DM to be supplied	DCP to be supplied	TDN to be supplied	Actual quantity of Feed/fodder to be supplied
DHN 6	2.22 kg	0.222 kg	1.32 kg	8.88 kg
Kadbi	4.44 kg	0 kg	1.86 kg	4.9 kg
Concentrate	3.33 kg	0.59 kg	2.33 kg	3.7 kg
Total		0.81kg	5.51 kg	

Salt and Mineral mixture to be supplied

Salt and mineral mixture to be supplied through ration=1% of the total quantity of concentrate mixture

Thus, salt and mineral mixture should be fed = 37 gm each/day

Example 4

Compute a ration for a buffalo weighing 500 kg and producing milk 10 liter with 7 % fat

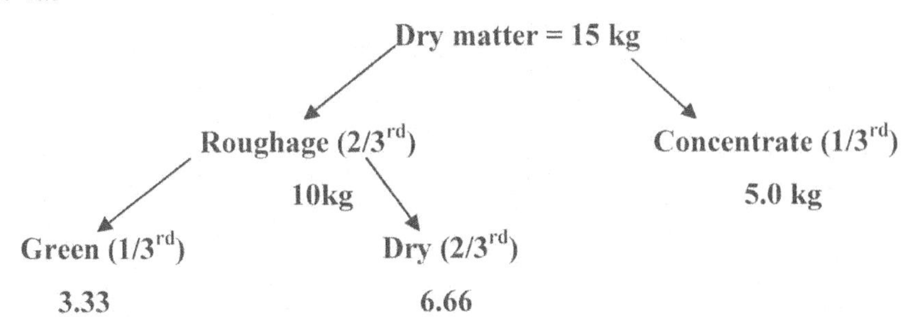

Available feed & fodder with composition

	DM	DCP	TDN
DHN 6	25%	10%	60%
Kadbi	90%	0%	42%
Concentrates	90%	18%	75%

Calculation of DCP and TDN requirement of animal

	DCP	TDN (for 10 ltr. milk)
Maintenance	0.290 kg	4.58 kg
Production	0.51 kg	3.69 kg
Total	0.806 kg	8.27 kg

Calculation of DCP and TDN to be supplied:

Feed/fodder	DM to be supplied	DCP to be supplied	TDN to be supplied	Actual quantity of Feed/fodder to be supplied
DHN 6	3.33 kg	0.33 kg	1.99 kg	13.32 kg
Kadbi	6.66 kg	0 kg	2.79 kg	7.4 kg
Concentrate	5.0 kg	0.9 kg	3.5 kg	5.55 kg
Total		1.233kg	8.28 kg	

Salt and Mineral mixtue to be supplied

Salt and mineral mixture to be supplied through ration=1% of the total quantity of concentrate mixture

Thus, salt and mineral mixture should be fed = 55 gm each/day

CHAPTER 12

MEASURES OF FEED ENERGY

- Energy is the capacity to do work
- Units of energy
    1. Calorie- It is the amount of heat required to raise the temparature of one gram of water from 14.5 to 15.5°C at atmospheric pressure.
    2. Kilocalorie = 1000 calories.

       Kilocalorie (Kcal) is the amount of heat required to raise the temparature of 1 kg of water to 10°C.

       Megacalorie (Mcal) or therm = 1000 kcal.

       1 Calorie = 4.185 J.

       The energy produced from feed is measured in Bomb calorimeter.

1. Gross energy (100%)

The total energy of the feed is called gross energy. It is also called as heat of combustion.

The gross energy value of a feed has no relationship to the feeds digestible, metabolisable, or net energy values, except that the latter can never exceeded the GE.

Certain products such as coal, mineral oil and lignin have high energy values but, because of their indigestibility are of no energy value to the animal.

Roughages have gross energy values comparable to concentrates, but the two differ greatly in digestible, metabolisable, and net energy values.

Fat, because of their greater proportion of carbon and hydrogen, yields 2.25 times more gross energy per kg than carbohydrates and protein.

A young growing animal will store energy principally in the protein of its tissues, fattening animal stores energy in fat and a lactating animal will transfer food energy into milk.

2. Digestible energy (70%)

Digestible energy is represented by that portion of feed energy consumed which is not excreted in the faeces.

It is the energy due to digested nutrients in the body.

3. Metabolisable energy (60%)

When the energy losses in the urine (3 to 5%) and combustible gases(8%) (Primarily methane (CH_4) are substracted from the digestible energy, the remaining energy is called the metabolisable energy.

Metabolisable energy = Energy in the food - (Energy lost in faeces+ energy lost in combustible gases + lost in urine)

Metabolisable energy can also be calculated from the digestible energy by multiplying with 0.82 which means roughly about 18 % of the energy is lost through urine and methane.

$$ME = DE \times 0.82$$

Factors affecting the metabolisable energy values of feeds

1. **Species of animals:** In the ruminants about 8-10 % losses of energy are in the methane production while in the non-ruminants there are no such losses.

   Therefore, the ME values are higher in non-ruminants than for the ruminants.
2. **Composition of feed**
3. **Processing of feed:** Processing of feed also affect the ME values since it affects the losses of nutrients in faeces and methane production.
4. **Level of feeding:** The level at which feed is being fed affect the ME value of the feed.

   At high level of intake ME values are reduced.

   Metabolisable energy is the useful energy of feed for the animal body.

4. Net energy (50%)

This is the net remainder of the useful energy after all losses accounted for faeces, urine, methane and heat increment is substracted.

This is the actual energy retained and utilised by the body for its growth and production.

It differes from metabolisable energy in that net energy does not include the heat of fermentation and nutrient metabolism or the heat increment.

CHAPTER 13

SYSTEMS OF BREEDING OF DAIRY ANIMALS

1. **Inbreeding:** Breeding of the related animals
2. **Outbreeding:** Breeding of the unrelated animals

Inbreeding

Involves mating of related individuals within 4 to 6 generations.

or

It is defined as the mating of the more closely related individuals than the avarage of the population.

Classification of inbreeding

1. Close breeding

   i.e. Sire to daughters

   or son to Dam

   or Full brother and sister

2. Line breeding

   i.e. Half brother and sister

   or mating of animals more distantly related

   e.g. cousin mating

Close breeding

In this type, if the both parents are having outstanding performance then & then only the parents are used for breeding.

Advantages of close breeding:

1. Undesirable recessive genes may be discovered and eliminated by further testing in this line.
2. The progeny are more uniform than outbred progeny.

Disadvantages of close breeding

1. The undesirable characteristics are infested in the progeny if unfavourable gene segregation occurs.
2. It has been observed that the progeny becomes more susceptible to diseases.
3. Breeding problems and reproductive failure usually increase.
4. It is difficult to find out the stage of breeding at which it should be discontinued in order to avoid the bad effects of the system.

Line breeding

- Mating of animals of wider degrees of relationship than close selected for close breeding is called as line breeding.
- It promotes uniformity in characters.
- In this method desirable or harmful characters are not developed so quickly.

Advantages of line breeding

1. Increased uniformity in characters.
2. The dangers involved in close breeding can be reduced.

Disadvantages of line breeding

1. Breeder will select the animal for pedigree without consideration of real individual merit. Therefore upto some generations no benefit from selection will be noticed.

Consequences in Inbreeding

Effect of both close breeding & line breeding are same. Only intensity is different between them.

1. It increases homozygocity(like allels) & decreases heterozygosity and hence favours the development of genetic uniformity amoungst animals.

Outward effects of inbreeding

1. Reduces growth rate in animals.
2. Reduced reproductive efficiency in the form of delay testicular development and puberty, reduced gametogenesis or increase in embryonic death.
3. Death rate is higher in inbred animals than outbred animals.

4. With increase in inbreding moderate decrease in production will be occurred.
5. Hereditary abnormalities or lethal factors are likely to appear more often in the inbred animals.

2. Out-breeding

Mating between unrelated animals is called as outbreeding

Types of outbreeding

1. Outcrossing
2. Cross-breeding
3. Species hybridization
4. Grading up

1. Outcrossing

- Mating of unrelated purebred animals within the same breed.
- The animals mated have no common ancestors on either side of their pedigree upto 4-6 generations.
- Offspring of such a mating is known as out-cross.

Outcrossing

Advantages:

1. This method is highly effective for characters that are largely under the control of genes with additive effects. e.g. milk production, growth rate etc.
2. It is effective system for genetic improvement if carefully combined with selection.
3. It is the best method for most herds.

Crossbreeding

- Mating between animals of different breeds is called as crossbreeding.
- Crossbreeding for milk production has been tried with varying degrees of success.
- Crossbreeding generally used for the production of new breeds.

Methods of cross breeding

1. **Criss-crossing:** When the two breeds are crossed alternatively, the method is called as criss-crossing.

   In this method breed A females are crossed with breed B sires. The cross-bred females are mated back to sires of breed A and so on.

2. **Triple cross:** Three breeds are crossed in a rotational manner. It is also known as rotational crossing.

   Three breeds are used in this system. The females of crosses are used on a sire of pure breeds in rotation.

3. **Back crossing:** Back crossing is mating of a crossbred animal back to one of the pure parent races which were used to produce it.

   It is commonly used for genetic studies, but not widely used by breeders.

   When one of the parents possesses all or most of the received traits, the back cross permits a surer analysis of the genetic situation than F2 does.

   A heterozygous individual of the F1 when crossed with a member of the heterozygous recesive parent race, the offsprings group themselves into a phenotypic ratio of 1:1.

Advantages of cross-breeding

1. Due to crossbreding desirable characters are introduced into breed in which they have not existed formerly.
2. Due to cross breeding new breed evolution is occurred.
3. It is an extremely handy tool to study the behaviour of characteristics in hereditary transmission.
4. The cross-bred animals usually exhibit an accelerated growth and vigour or heterosis.

Disadvantages of crossbreeding

1. Crossbreeding has a tendency to break up established characters and destroy combination of characters which have long existed in the strains.
2. Crossbreding requires maintenence of two or more pure breeds in order to produce crossbreeds that will become more expensive.

Species hybridization, Heterosis or hybrid vigour

- The phenomenon in which the crosses of unrelated individuals often results in progeny with increased vigour much above the parents.

Or

- The increased level of performance as compared to the avearge of the parental types is known as heterosis.
- The progeny may be from the crossing of strains, breeds, varieties or species
- The increased vigour is due to that genes favourable to reproduction are usually dominant over their opposites. When one breed is crossed with the

other, one parent supplies a favorable dominant gene to offset the recessive one supplied by the other and vice versa.

- The offspring, therefore, has a larger number of loci with dominant genotypes than does either parent and is likely to be more vigourous.
- The increased vigour is due to overdominance, where the heterozygous condition is superior to either homozygous condition.
- Heterosis is employed to produce commercial stocks with high individual merit.
- It is generally used in poultry, pigs & sheeps.

Grading up

- Grading up is the practice of breding sires of a given breed to non-descript females and their offspring for generation after generation.
- It is the succesive use of purebred bulls of a certain breed of non-purebred herds.
- The continued use of good purebred sires for only few generations are all that are required to bring the herd to the point at which it has all the appearance, actions and practical values of pure breeds.

Advantages of grading up

1. Purebreds can be obtained just after a few generations(After 7 to 8th generations).
2. The start can be made with a little money in comparison to the purchase of an entire herd of purebreds.
3. It helps to prove the potentialities of the sire and adds to its market value.
4. It is a good start for new breeders who can slowly change over to pure breed systems.

Limitations of grading up

1. Pure breeds are not always better than grade or country animals for the use to be made of them.

   Purebreed stock which give good results in one set of environmental conditions do not always give favourable results in some different environmental setup. The purebred dairy cattle from temporate zones often degenerate when used in tropical areas.

CHAPTER 14

DAIRY FARM RECORDS AND THEIR MAINTENANCE

The business without proper record is impossible. Unless accurate record is kept, the best cow in a herd is likely to have equal rank with poorest atleast in the mind of owner.

There are 3 methods of preservation of records, these are,

1) In books with permannent leave
2) In loose leaf books are files
3) Envelops

Various records of the farm activities and business are maintained in different register. These are as follows,

1. Livestock register
2. Calf register
3. Milk record register
4. Feed and fodder register
5. Herd health register
6. History and pedigree sheet register
7. Growth register for young stock
8. Mortality register
9. Cattle breeding/ Events register
10. Finanacial account register
11. Daily livestock account
12. Calving register
13. Lactation/Milk yield register

Proforma of different dairy farm records

1. Service/Event register

Sr.No.	No. of cows	Date of last calving	Date of service	Time of service	No. of bull	Result of service	Expected date of calving	Date to be dried off	Date of calving	Weight of calf	Time taken for the expulsion of placenta	Remarks

2. Feed stock register

Sr. No.	Bill no.& date	From whom it purcahsed	Quantity	Rate	Amount (Rs.)	To whom it issued	Date of issue	Indent no.	Remarks	Siganature of officer

3. Feed utilization/ daily ration register

Date	Opening balance(kg)	Received feed in kg	Ration fed to animals	Particulars	Signature of officer

4. Calf register/Youngstock register

Sr.No.	Date of birth	Date of numbering	Tag no.	Sex of calf	Sire	Dam	Birth weight of calf	Disposal		Remarks
								How the calf disposed	Date	

5. Milk record register

Date	Time	Name and number of animal	Name and number of animal	Name and number of animal	Name and number of animal
1	A.M				
	P.M.				
10	A.M.				
	P.M.				
30	A.M.				
	P.M.				

6. Health record of individual animal

Name of farm:

Type of animal: Breed: Sex: No.

Tuberculosis / Brucellosis/ JD tests		General health				
Date	Results					

7. Lactation record

Animal No.	Months of year						Yield in lit. during lactation	Average fat%	No. in days in a lactation	Date and days dry off	Remarks
	Jan.	Feb.	.....	....	Nov.	Dec.					

8. Stock register of animal

Sr. no.	Tag no.	Date of purcahse	Date of birth	Value	From whom it received	Age at purchased	Pedigree		How disposed off	Page of herd register	Remarks
							Dam	Sire			

Objectives for keeping the record on dairy farm are

1. It helps to access the present status of dairy business.
2. It helps in perspective planning of business.
3. It helps in management of farm efficiently.
4. Individuals which are low producer can be identified at early stage of lactation.
5. It helps to select animals for culling.
6. Profitability and loss can be judged at any time of year.
7. It helps in implementation of systematic breeding policy.

CHAPTER 15

COMMON DISEASE PROBLEMS OF DAIRY ANIMALS, THEIR PREVENTION AND CONTROL

Diseases of dairy animals

1. Calving disorders
2. Early lactation pitfalls
3. Everyday problems
4. Infectious diseases

Calving disorders

1. Dystocia
2. Paralysis
3. Retaintion of placenta
4. Prolapsed Uterus
5. Metritis
6. Milk fever

1. Dystocia

- Difficulty in calving is called dystocia.
- No progress after calf is in birth canal.

Causes

- Calf not presented properly
- Large calf
- Twins
- Monsters

Diagnosis

- Vaginal examination
- Must recognize presentation
- Front-2 legs & head
- Rear -2 rear legs
- Spine to spine
- Always clean hands before examination.

Treatment

- Assisted pulling.
- Correct problems in presentation.
- Consider Veterinarians help-especially if something feels unusal.

Prevention

1. Breed at proper age & size.
2. Provide adequate nutrition to pregnant animals.
3. Give adequate exercise during pregnancy.
4. Avoid long distance transportation, especially during advance pregnancy.
5. Avoid rolling, struggling, falling & jumping in pregnant animals.
6. Culling of animals with known pelvic deformities or kinked cervix.

2. Calving paralysis

- Damage to nerve & muscle due to trauma during calving.
- Reason: Calf not pulled properly.
- Dystocia.

Diagnosis:

- H/O hard calving
- Cow usually alert, eating, drinking
- Normal temparature

Treatment

- Tincture of time
- Lifting
- Nervine tonics
- Feeding with protein diet

3. Retaintion of placenta

- Failure of expulsion of placenta even after 12 hours of parturition.

Etiology

- Abortion
- Dystocia
- Premature delivery
- Lack of exercise
- Deficiency of hormones like oxytocin/ estrogen
- Infectious diseases like Brucellosis, Vibriosis, Trichomoniasis
- Nutritional deficiency of calcium & phosphorus.

Symptoms

- Placenta retains even after 12 hours of parturition
- Straining
- Foul smelling discharge
- Fever
- Loss of appetite
- Dullness and depression

Treatment

- Remove placenta manually.
- Give injection Oxytocin IM at approximately 12 hours after parturition.
- Keep 2-4 furea boluses intrauterine for 1-2 days.
- Give uterine cleansers.
- Parentral administration of antibiotics.

4. Prolapsed uterus

- Eversion of genital organs is called uterine prolapse.

Etiology

- Inherited tendency
- Low levels of progesterone
- Urogenital infections like cervicitis & vaginitis
- Dystocia
- Breeding injury
- ROP
- Straining due to diarrhoea or constipation

Symptoms

- Protrusion of uterus, cervix and/or vagina beyond vulva
- Continuous straining
- Wounds or injury on the prolapsed mass
- Restlessness
- Rise in body temparature

Treatment

- Washing of prolapsed mass with antiseptic solution.
- Application of antiseptic ointment on prolapsed mass.
- Reduction of protruded part by cold fomentation.
- Reposition of the prolapsed mass manually.
- Apply rope-truss to provide support and for retaintion.
- Keep the animal in slanting with hindlegs at higher level & head at lower level.
- In antepartum prolapse, give injections of progesterone on every 10th day.
- Intrauterine and parentral antibiotic therapy.

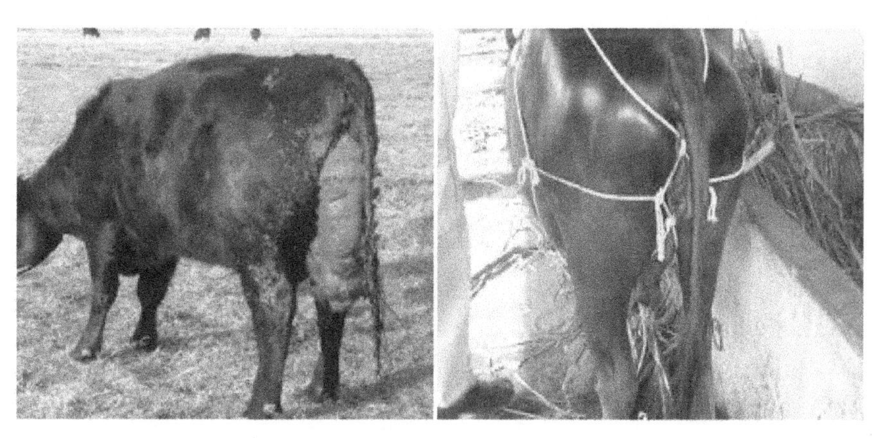

Prevention

- Eliminate causes of irritation or straining.
- Avoid injury or unnecessary traction at delivery.
- Early treatment of ROP.
- Don't apply weight to hanging placenta to remove it.
- Proper feeding during gestation.

5. Metritis

- Inflammation of uterus is called metritis.

Etiology

- Bacteria: Brucella, Staphylococci, Streptococci
- Protozoa: Trichomonas foetus
- Dystocia
- ROP
- Abortion
- Delayed involution of uterus
- Uterine inertia
- Lack of exercise
- Injury during insemination
- Unhygienic conditions at the time of parturition

Symptoms

- Rise in body temparature
- Rapid pulse
- Purulent discharge from uterus
- Drop in milk yield
- Reduction in body weight
- Straining
- Frequent urination
- Loss of appetite

Treatment

- First remove uterine discharge by gentle per-rectal uterine massage
- Give antibiotics for 3-5 days
- Intra-uterine therapy for 2-4 days
- Inj. E care Se

6. Milk fever

- It is metabolic disease of adult females.

Etiology:

- Hypocalcemia due to
    1. Excessive loss of calcium in the colostrum and milk.
    2. Decreased absorption of calcium from the intestine at the time of parturition.
- Reduced feed intake
- Ruminal atony
- Deficiency of vitamin D
- Improper calcium-Phosphorus ratio in diet
- Diseases of intestine

    3. Slow mobilization of calcium from bone
- Parathormone deficiency
- High calcium level in blood
- Excessive calcium intake during dry period depresses the activity of PTH

Clinical signs

- **First stage:**

    Excitement
    - Tetany
    - Hypersensitiveness
    - Muscle tremur
    - Ataxia
- **Second stage:**

    Animal unable to stand & rests on sternum with head turned towards shoulder or flank.

- o Drowsiness
- o Cold skin, extremities
- o Low body temparature
- o Dilation of pupil
- o Heart sounds muffled

Treatment

- Parentral injection of calcium salt
- Calcium borogluconate 25% @ 400-800 ml IV
  Inj. Vetade 5 ml IM

1. Early lactation pitfalls

1. Abomasal displacement
2. Ketosis
3. Rumen acidosis
4. Rumen ulcers
5. Fatty liver

1. Abomasal displacement

Etiology

- Poor exercise
- Heavy feeding of concentrate
- Hypocalcemia
- Ketosis
- Metritis
- Act of calving

Clinical signs & symptoms

- H/O recent calving
- Anorexia
- Drop in milk production
- Ketonuria
- Colic
- Scanty pasty faeces
- Buldging of lower left paralumbar fossa
- Ascultation yield ping sound

Treatment

- Rolling & manipulation
- Surgical correction-
  1) left flank omentopexy
  2) Left & right flank abomasopexy
  3) Right flank omentopexy
  4) Ventral paramedian abomasopexy

2. Ketosis

- It is metabolic disease of high yielding animals characterised by hypoglycemia, ketonaemia and ketonuria.

Clssification

1. **Primary ketosis:** It indicates starvation or underfed or uncomplicated ketosis.
2. **Secondary ketosis**: caused by various systemic or infctious diseases.

Etiology

- Undernutrition/starvation or feeding low carbohydrate diet
- Feeding of excess amount of protein-rich diet
- Excess feeding of silage
- High milk yield
- Deficiency of cobalt and phosphorus
- Lack of exercise
- Hepatic insufficiency
- Adreno-cortical deficiency
- Hypothyroidism
- Loss of appetite due to various diseases

Symptoms

- Selective appetite: Animal refuses to eat concentrate feed however takes roughages
- Marked drop in milk production
- Rapid loss of body weight
- Moderate depression and disinclination to move
- Sweetish smell to breath, milk & urine
- Ruminal movements may be decreases in amplitude & rate

Treatment

- **Specific treatment:**
  1. Glucose therapy alone: Glucose 20% @ 0.5g/kg IV daily for 2-3 days.
  2. Glucose+insulin therapy: Glucose 20% @ 0.5g/kg IV plus insulin(Short acting) @ 0.5 IU/kg IM only once.
- Corticosteroids

Supportive treatemnt:

- Inj. Liver extract with B-complex @ 5-10 ml IM on alternate day
- Provision of mineral mixture comprising phosphorus and cobalt

3. Everyday problems

1. Mastitis

Etiology:

A. Infectious agents

Bacteria: Streptococcus, Staphylococcus, E.coli

Viral diseases: Cow pox, FMD

Fungus: Aspergillus, Candida, Cryptococcus

Mycoplasma

B. Predisposing factors

Trauma or injury to teat and udder

High milk yield

Incomplete or irregular milking

Improper milking techniques

Pendulous udder & long cylindrical teats

Rough flooring

Unhygienic conditions

Transmission

It spreads through

- Infected water,
- Contaminated bedding
- Utensils
- Milker hands

Symptoms

A. Acute form:

1. Fever
2. Loss of appetite
3. Udder swollen, hot and painful
4. Milk may be yellowish or brownish
5. Milk contains flakes or clots

B. Chronic form

1. No swelling of udder
2. Udder becomes hard due to fibrosis
3. Milk may show visible changes on careful examination
4. Reduced milk yield

Diagnosis

- Physical examination of udder
- Strip cup test
- CMT Test
- Isolation of the organism from milk

Treatment

1. Evacuate the udder
2. Intramammary antibiotic therapy
3. Parentral antibiotics
4. Hot fomentation of udder
5. Antiinflammatory, analgesic therapy

2. Pneumonia

- Inflammation of lungs is called as Pneumonia.

Etiology

- Bacteria
- Viruses
- Fungi
- Parasites
- Allergens
- Faulty drenching
- Stress

Clinical signs & symptoms

- Fever
- Anorexia
- Dyspnoea
- Coughing
- Nasal discharge
- Rales on ascultation of chest

Treatement

- Rest
- Warmth
- Antibiotics
- Expectorants
- Antiinflammatory & analgesics
- Antihistaminics
- Bronchodilators
- Anthelmintics

3. Enteritis(Diarrhoea)

Inflammation of intestinal mucous membranes is called as enteritis.

It may be chronic or acute.

Etiology

Dietary

- Overfeeding
- Sudden change in diet
- Coarse/fibrous feed
- Green lush pasture, green grass
- Mouldy feed
- Decomposed/spoiled feed
- Excess dietary carbohydrates
- Excess feeding of milk or milk replacer
- Ingestion of poisonus plants

Physical agents

- Sand & soil

- Chemical agents

    Irritant chemicals: Cu, Flourine, Mercury, Nitrate

    Drugs: Antimicrobials, Anthelmintics

- Bacterial agents

    Enterotoxaemia and JD

- Viral agents

    BVD, BMC, Rinderpest, Rotavirus, Corona virus

- Protozoan agents

    Coccidiosis, Balantidiasis, Cryptosporidiasis

- Fungal agents

    Candida, Aspergillus

- Parasites

    Roundworms: Ostertagia, Trichostrongylus, Oesophagostomum, Bunostomum, Trichuris, Cooperia

- Tapeworms: Moniezia spp.
- Flukes: Fasciola hepatica, F. gigantica
- Nutritional deficiency
- Deficiency of copper,Cobalt, Vit.A

Symptoms

A) Acute

Diarrhoea/Dysentry

Faeces are soft or fluid in consistency, foul or fishy odour, abormal colour

Frequency of defecation increased

Soiling of buttock & hind quarters

Straining

Dehydration: Sunken eyes, rough coat, tenting of skin

Oligourea

B) Chronic:

Faeces usually soft & homogenous in consistency

Faeces may contain considerable mucus

Faeces usually do not have abnormal odour

Emaciation

Dehydration is not marked

Treatment

1. Removal of causative agent
2. Antibacterial drugs oral/ parentral
3. Astringents
4. Fluid or electrolyte therapy
5. Antispasmodic to control colic & straining
6. Alteration in diet

4. Infectious diseases

1. Bacterial diseases:
    1. Haemorrhagic Septicaemia
    2. Black Quarter
2. Viral diseases:
    1. FMD

1. Bacterial diseases:

1. Haemorrhagic Septicaemia

   It generally occurs in rainy season.

Etiology

- Pateurella multocida

Transmission

- Ingestion of contaminated feed & water
- inhalation

Symptoms

- High fever(106-107oF)
- Loss of appetite
- Suspended rumination
- Profuse nasal discharge
- Difficult/snoring respiration
- Swelling of throat region(submandibular edema
- Shivering
- Recumbency & death within 10-72 hrs

Treatment

1. Specific treatment

Inj. Enrofloxacin/Sulphadimidine/Oxytetracycline

- **Supportive treatment**
    1. Use of antipyretics
    2. Use of antihistaminics
    3. Fluid therapy

2. Black Quarter

- Young cattle between 6-24 months of age, in good body condition are mostly affected.
- It is soil-borne infection.
- It is generally ocurrs during rainy season.

Etiology

- Clostridium chauvoei

Transmission

- Ingestion of contaminated feed
- Contamination of wounds

Symptoms

- Fever(106-108°F)
- Lameness of affected leg.
- Crepitating swelling over hip, back & shoulder.
- Swelling is hot & painfull in early stages whereas cold & painless later.
- Rapid pulse & heart rate.
- Difficult breathing.
- Loss of appetite.

Treatment

1. Antibiotic therapy:Strepto-penicillin
2. Incise the swelling & drain off
3. BQ antiseum in large doses, if available
4. Use of antipyretics
5. Use of antihistaminics
6. Fluid therapy

2. Viral diseases

Foot and Mouth disease

- It is highly contagious viral disease of cloven footed animals.
- The disease usually occurs at the end of winter i.e. February, March.

Etiology

- Picorna virus(Apthovirus)

Transmission

- ingestion of contaminated feed & water
- Air-born infection
- Suckling calves may pick up infection from dam/mother

FMD-Symptoms

- Fever(104-106°F) for 24-48 hrs
- Blister/vesicles and ulcers on toungue, dental pad & oral mucosa
- Profuse salivation
- Painful mastication
- Loss of appetite
- Vesicles & ulcers develop in interdigital space & on the coronet
- Lameness

Post complications of FMD

- Abortion
- Infertility
- Mastitis
- Pneumonia
- Deformity in hooves
- Anemia
- Excess hair growth
- Panting

Treatment

- Antibiotic therapy
- Mouth wash with 1% $KMnO_4$ or 2% Sodium biocarbonate solution
- Apply boroglycerine on mouth lesions
- Foot wash with 2% Copper sulphate or 2-4 % Sodium carbonate solution
- Apply mixture of coal tar and copper sulphate(5:1) on foot lesions

General preventive & control measures of diseases

1. Animals are fed with balanced diet.
2. Avoid sudden change in diet.

3. Avoid feeding kitchen waste.
4. Keep animal shed clean & dry.
5. Identification of isolation of infected & incontact animals.
6. Treatment of affected animals.
8. Slaughter of animals suffering from incurable diseases.
9. Vaccination of animals.
10. Regular deworming & ectoparasite control.
11. Disposal of dead animals either by burning or deep burial.
12. Destroy contaminated fodder by burning.
13. Regular disinfection of cattle shed & its premises with 1-2 % phenyl.
14. Don't allow grazing in affected area.
15. Restrict the movement of animals from affected to clean area.
16. Provide adequate ventilation & sunlight.
17. Avoid overcrowding.

CHAPTER 16

DIGESTIVE SYSTEM OF CATTLE/ BUFFALOES

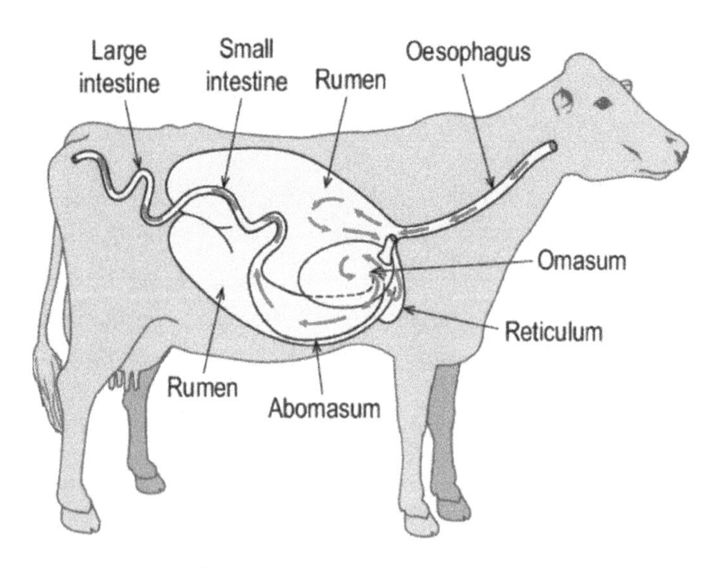

A] Alimentary canal

i] **Mouth [Lips and mouth cavity]:** It is the organ of prehension, mastication, insalivation and rumination.

ii] **Pharynx:** Its muscular walls help in the process of swallowing and it serves as a pathway for the movement of food from the mouth to oesophagus.

iii] **Esophagus:** It is the direct continuation of the pharynx to rumen. The feed bolus passes from pharynx to rumen through esophagus.

iv] **Rumen [Absent in non-ruminants]:** It is the main site for microbial digestion in ruminants. It also stores the eaten feed & fodder for rumination.

v] **Reticulum [Absent in non-ruminants]:** It collects smaller digesta particles and move them into the omasum, while larger particles remain in the rumen for further digestion. Reticulum also traps and collects heavy/dense objects the animal consumes.

vi] **Omasum [Absent in non-ruminants]:** Absorption of water & volatile fatty acids.

vii] **Abomasum [Stomach in non-ruminants]:** Abomasal cells secrete electrolytes, specially HCL, Pepsin and mucus. It provides optimum conditions for activity of the peptic enzymes responsible for the digestion of microbial protein in the abomasum.

viii] **Small Intestine:** The small intestine is the chief site of absorption of nutrients. Digestion of fat occurs in small intestine. Duodenum, Jejunum and Ileum are the parts of small intestine.

ix] **Large intestine:** It removes water from ingesta and absorbs in the body. Cecum/ caecum [Plural- ceca/caeca] Colon and rectum are the three parts of large intestine.

x] **Rectum & anus**

B] Accessory Digestive Organs

i] **Teeth:** It is useful for mastication of feed & fodder.

ii] **Tongue:** It is useful for rumination & in salivation of feed.

iii] **Salivary glands:** Secretes saliva useful for digestion & maintenance of Ruminal pH.

iv] **Liver:** Bile secretion occurs in liver.

v] **Pancreas:** Secretes pancreatic juice useful for protein digestion.

CHAPTER 17

MALE REPRODUCTIVE SYSTEM

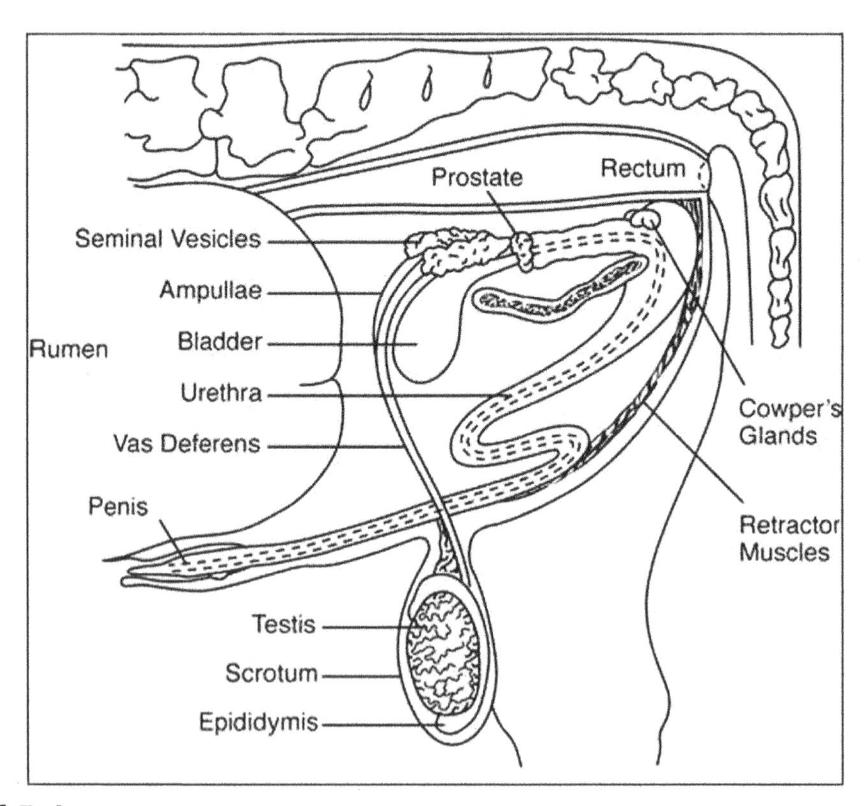

A] Primary sex organ

I] Two testes/testicles suspended in scrotum [Singular - Testis/Testicle]. The hormone testosterone is produced in testicles which is the sex hormone of male. Also sperms are produced in testis of male.

b] Secondary sex organs

i] **Vas efferentia**

ii] **Epididymis:** Accumulation of spermatozoa & maturation of spermatozoa occurs in Epididymis.

iii] **Vas deferentia:** It is involved in ejaculation of semen.

iv] **Penis:** It is the copulatory organ of male.

C] Accessory sex organ

i] **Prostate gland:** The secretion from prostate gland is high in mineral content.

ii] **Two seminal vesicles:** The secretion from seminal vesicle contains high level of fructose (upto 1%) and of citric acid.

iii] **Two bulbo-urethral glands or Cowper's gland:** Secretion is viscid or mucous type.

CHAPTER 18

FEMALE REPRODUCTIVE SYSTEM

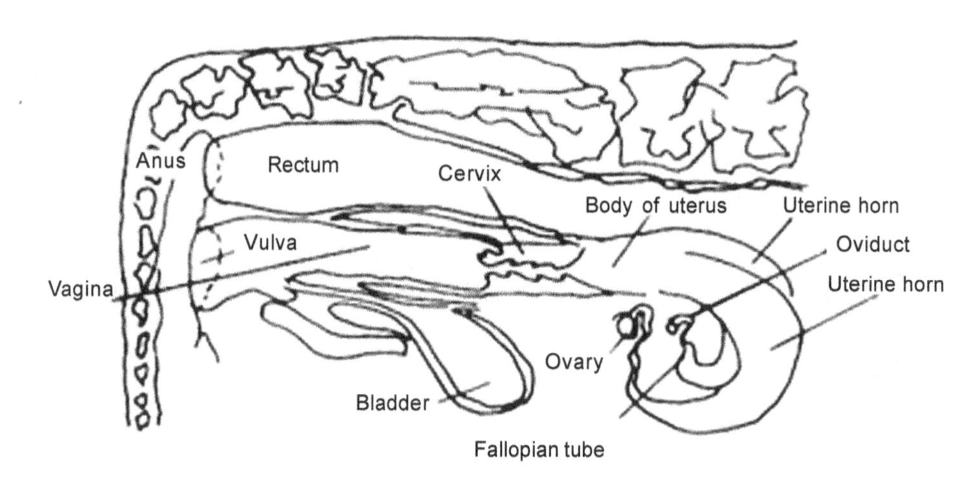

i] **Two ovaries:** Function: production of ova & endocrine function.

ii] **The uterine or fallopian tubes:** These tubes help in conducting the ova from the ovaries to the uterus.

iii] **Uterus & two uterine horns:**

Function: Maintenance of calf upto birth.

iv] **Cervix:** It is the part between vagina & uterus. It is closed during pregnancy & opens during parturition & oestrus.

v] **Vagina:** It is the organ of copulation in females, extending from the cervix posterior up to vestibule. The vagina serves as a birth canal at the time of parturition and admits the male organ in copulation.

vi] **Vulva:** It is the external vertical opening of the genital tract and lies just below the anus.

vii] **Clitoris:**

CHAPTER 19

ESTROUS CYCLE IN COWS / BUFFALOES

- It is appears in the normal female in breeding season
- Four phases of estrus cycle
  1. Proestrum
  2. Estrus or estrum
  3. Metoestrun or metestrum
  4. Dioestrum or Diestrum

1. Proestrus

- During this phase the graffian follicle within the ovary is growing with increase in secretion of follicular fluid.
- The follicular fluid surrounds the ovum, contains hormone estradiol
- The hormone estradiol absobed in the blood & it passes to oviduct or fallopian tube.
- The hormone estradiol causes growth of cells in oviduct or fallopian tube & increases the number of cilia (required for transpotation of ova to the uterus).
- In this phase, vascularity of the uterine mucosa is increased.
- The epithelial wall of vagina increases in thickness, so as to avoid injury during coitus.

2. Estrus

- This is the period of sexual desire by female

- The graffian follicles now 'riped' or become very turgid and the ovum is undergoing maturation changes.
- Near to end of estrus the ovulation may occurs
- Estrus period in cow lasts from 12 to 24 hrs.
- The vulva becomes swollen & both vulva and vagina are congested with blood

Symptoms of heat in cows & buffaloes

1. Smelling other cows
2. Attempting to mount on other cows & bellowing
3. Becomes restlessness
4. Vagina gets moist, red, & slightly swollen
5. The cow stands still to be mounted by other cows or bulls. This period is called as standing heat.
6. Nervousness & reduction in feed intake & milk yield
7. Clear & ropy vaginal mucous discharge
8. Frequent urination
9. On perrectal palpation uterus found tight & turgid, cervix open

3. Metoestrum or Metestrum

- In this stage the reproductive organs of females returns to normal non-congested condition.
- In metoestrum period, the cavity of the graffian follicle from which ovum has been expelled becomes reorganised and forms a new structure known as corpus luteum(CL), an endocrine gland which secretes progesterone.

Functions of progsterone

1. It prevents the maturation of further graffian follicles & thus prevents the occurance of further estrus period for a time.
2. It is essential for the implantation of fertilised egg.
3. It nourishes the foetus during the first half of pregnancy.
4. It is intimately concerned with the development of the mammary gland.

4. Dioestrum or Diestrum

- This is the longest part of estrus cycle.

- The CL is fully grown during this phase.
- The muscles of the uterus developed.
- If pregnancy supervenes, the uterus produces a copious supply of uterine milk for the nourishment of embryo.
- During gestation the CL remaining intact for whole pregnancy period.
- In absence of fertilised egg, the CL regresses completely.

CHAPTER 20

ARTIFICIAL INSEMINATION AND ITS ADVANTAGES

Artificial insemination

- Deposition of male reproductive cells i.e. sperms in female reproductive tract by artificial means is called as artificial insemination.
- **Liquid nitrogen container**

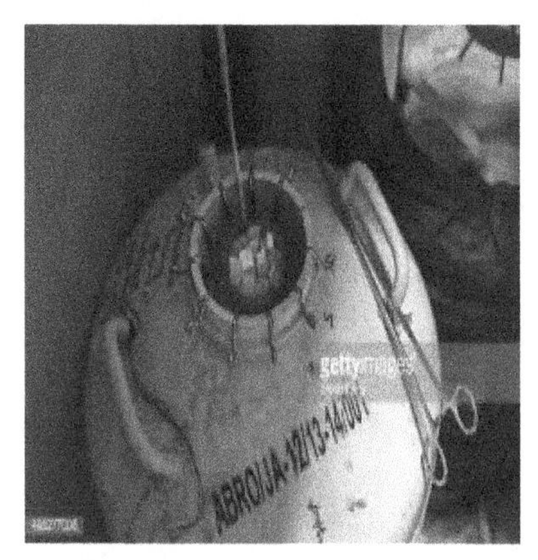

Artificial insemination technique

1] Examination of cow for proper stage of oestrus [Use of vaginal speculum]

2] Semen straw to load in AI gun by following thawing procedure at 37°C.

3] Wear full sleeve gloves on left hand, lubrication by soap and water. Trimming of nails of the fingers is necessary to avoid injury to rectal mucosa.

4] Grasping of cervix of cow after removal of faeces from her rectum.

5] Gun to be inserted into cervix very gently.

6] Deposition of semen in the middle cervix or uterus by pressing the piston of the gun.

7] Withdrawal of the AI gun gently.

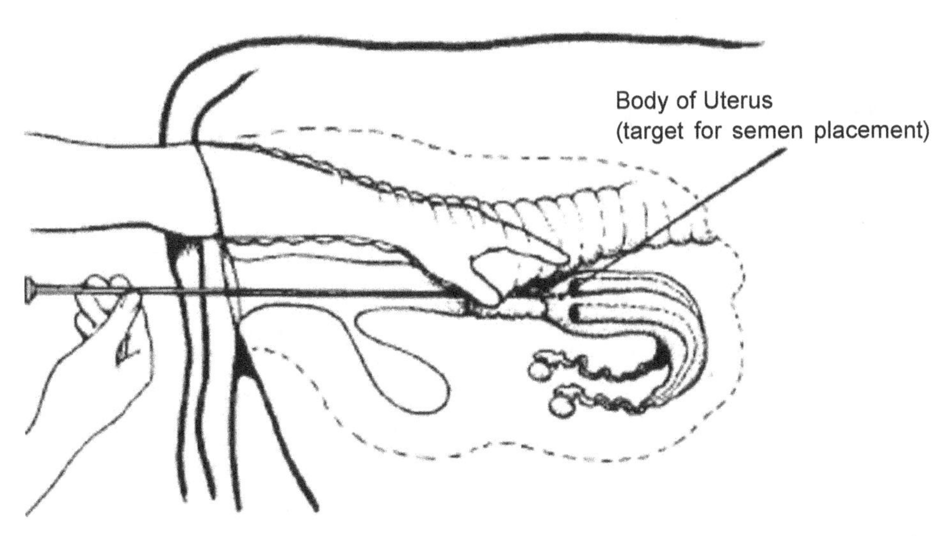

Advantages of AI

1] Increased use of bull for more number of cows.

2] Extended period of reproductive life of valuable sire.

3] Cost of maintenance on bull is saved.

4] Cost of transportation is also curtailed many folds.

5] Semen of heavy or injured bull can also be used.

6] Semen can be transported to remote areas where bull cannot be made available.

7] Helps proper record keeping.

8) Prevents spread of contagious diseases from one bull to females during natural service.

Limitations of AI

1] Costly equipments are required.

2] Cleaning, sterilization of equipments is necessary.

3] Strict maintenance of personal hygiene.

4] Trained and experienced staff is needed.

CHAPTER 21

NUTRIENTS AND FERTILITY IN ANIMALS

Repeat breeding in animals is due to abnormalities in ovulation, early embryonic mortality, anovulation, oviduct obstruction, Abnormal Environment of Oviducts and Uterus:, Semen and Incompatibility, Management Problems, Genetic or Congenital Anomalies of the Genital Tracts, nutritional deficiencies.

Animal nutrition plays important role in animal reproduction. The conception of the cows is associated with body weight. Mandatory weight which cows should achieve before breeding is for indigenous and jersey cross heifer 240-275 kg and for HF cross heifer 260-290 kg. Underweight animals show poor rates of conception. Use of milk replacer, calf starter and cereal improved fodders and leguminous fodders for the feeding of calves definitely leads to attend mandatory body weight before breeding. Providing balanced ration to calves will results into attending the early maturity body weight in animals.

A frequently used mechanism for sperm–egg recognition in many species involves complementary protein–carbohydrate interaction. The usual paradigm includes complex glycoconjugates in reproductive tract fluids or on the eggs which are recognized by carbohydrate binding proteins on the sperm surface. Various glycocojugates are utilized in the steps of sperm capacitation, sperm binding to the egg extracellular matrix and vitelline membrane and induction of the acrosome reaction. Several types of complex glycoconjugates are involved in these processes, including proteoglycans, lactosaminoglycans, sulfated fructose containing glycoconjugates, and glycoproteins. Hence carbohydrates plays important role in fertilization of eggs.

Cows fed excess protein (more than 10-15% above requirements) required more services per conception and had longer calving intervals. Feeding excess

RDP (rumen degradable proteins) has a negative effect on fertility. It clearly states that feeding high levels of RDP delays the first ovulation or oestrus, reduces the conception rate to first insemination, increases the number of days open and lowers the overall conception rate. There are several proposed mechanisms for this effect including an exacerbated negative energy balance for cows fed diets high in RDP in comparison to diets high in RUP and proven deleterious effects of both ammonia and urea on both oocyte and embryo development. However, this deleterious effect of excess RDP ingestion may be absent in beef heifers, which are normally in positive energy balance at breeding. Dietary excesses of metabolisable proteins (MP) are thought to be of lesser consequence than excesses of RDP in terms of dairy cow fertility, excess MP supply relative to requirement will often increase milk yield and exacerbate negative energy balance. Hence supply relatively low amounts of MP, in order to reduce milk yield response to energy supplementation and thus improve energy balance. Hence it is essential to use 35% of the crude protein as an undegradable protein in high producing and early lactation cows.

Fats in the diet can influence reproduction positively by altering both ovarian follicle and corpus luteum function via improved energy status and by increasing precursors for the synthesis of reproductive hormones such as steroids and prostaglandins. Dietary fatty acids of the n-3 family reduce ovarian and endometrial synthesis of prostaglandin F_2 alpha, decrease ovulation rate and delay parturition in animals.

Phosphorus has been most commonly associated with decreased reproductive performance in dairy cows. Inactive ovaries, delayed sexual maturity and low conception rates have been reported when phosphorus intakes are low. In a field study when heifers received only 70-80% of their phosphorus requirements and serum phosphorus levels were low, fertility was impaired (3.7 services per conception). Services per conception were reduced to 1.3 after adequate phosphorus was supplemented.

Sodium and Potassium are indirectly related to reproduction in animals as the deficiency of sodium can affect the normal reproductive physiology by preventing the utilization of protein and energy where as deficiency of potassium is well known to cause muscular weakness and thereby affect the musculature of female genital tract causing impairment in the normal reproductive process. It was also noticed that feeding high levels of potassium (5% DM basis) may delay the onset of puberty, delay ovulation, impair corpus luteum (yellow body) development and increase the incidence of anestrus in heifers. Lower fertility was noticed in cows fed high levels of potassium or diets in which potassium-sodium ratio was too wide.

Copper is one of the important mineral for reproduction point of view as its deficiency may leads to early embryonic death and resorption of the embryo. In addition to this, proper copper supplementation is must for quality semen production.

The reproductive processes affected due to molybdenum deficiency are decreased libido, reduced spermatogenesis, and sterility in males and delayed puberty, reduced conception rate in females.

Zinc is essential to exhibit secondary sexual characteristics and development of gonadal cells in males. Deficiency of Zinc in male's leads to poor semen quality, reduced testicular size and libido. In females Zinc is essential to maintain and repair of uterine linings.

Deficiency of selenium may results into weak, silent or irregular estrus, early embryonic death, still birth or weak offspring and abortions in females.

Low level of Manganese in blood of animals leads silent estrus and anoestrus (Corrah, 1996) or irregular estrus and decrease conception rate, birth of deformed calves and abortions in females and absence of libido and improper or failure of spermatogenesis in males. Manganese is important in cholesterol synthesis which in turn is necessary for the synthesis of steroids like progesterone, estrogen and testosterone. Decrease concentration of these steroids in circulation following manganese deficiency may lead to related reproductive abnormality.

Delay in onset of puberty, delay uterine involution and decreased conception rate are observed in case of Cobalt deficiency in animals.

Signs of iodine deficiency include delay in puberty, suppressed or irregular estrus, failure of fertilization, early embryonic death, still birth with weak calves, abortion in females and decrease in libido and deterioration of semen quality in males. Reproduction is influenced through iodine's action on the thyroid gland. Inadequate thyroid function reduces conception rate and ovarian activity.

Chromium exerts a significance influence on follicular maturation and luteinizing hormone release and thereby affects fertility in animals.

Therefore to avoid repeat breeding in animals balanced feed along with mineral mixture is proved to be effective.

CHAPTER 22

OVULATION, FERTILIZATION, GESTATION, PREGNANCY DIAGNOSIS AND PARTURITION

Ovulation

- It is defined as the discharge of the egg from the graffian follicle
- In cows & buffaloes ovulation occurs 12-15 hours of end of estrus
- In most domestic animals, ovulation occurs whether or not mating has occurred. In these animals the ovulation depends on blood level of FSH and LH.

Mechanism of ovulation

1. The ovulation is occurs under the control of LH. The mature follicle wall consists of three layers. The outer layers separate during final pre-ovulatory changes and the inner layer protrudes forming papilla. Finally inner layer gives way and the ovum with attached cells flow out.
2. Released LH stimulates DNA to synthesis messenger RNA (m-RNA). This inturn aids for the synthesis of the enzyme-collagenases. Collagenases in turn causes enzymatic decomposition of the collagen content of the follicular wall structure resulting in the oozing process of ovulation
   - In domestic animals the life of an egg varies from 12 to 24 hrs after ovulation.
   - Ovulation occurs once a month in bovine species.
   - Ovulation in the cow occurs more frequently in the right ovary than in the left.

Fertilization

- Penetration of the comparatively large sessile (stalkless) egg by a small motile spematozoan is called as fertilization of egg.

Capacitation of sperms

- The changes within sperm that confer upon the sperm the ability to acrosome react in response to the appropriate stimulus.

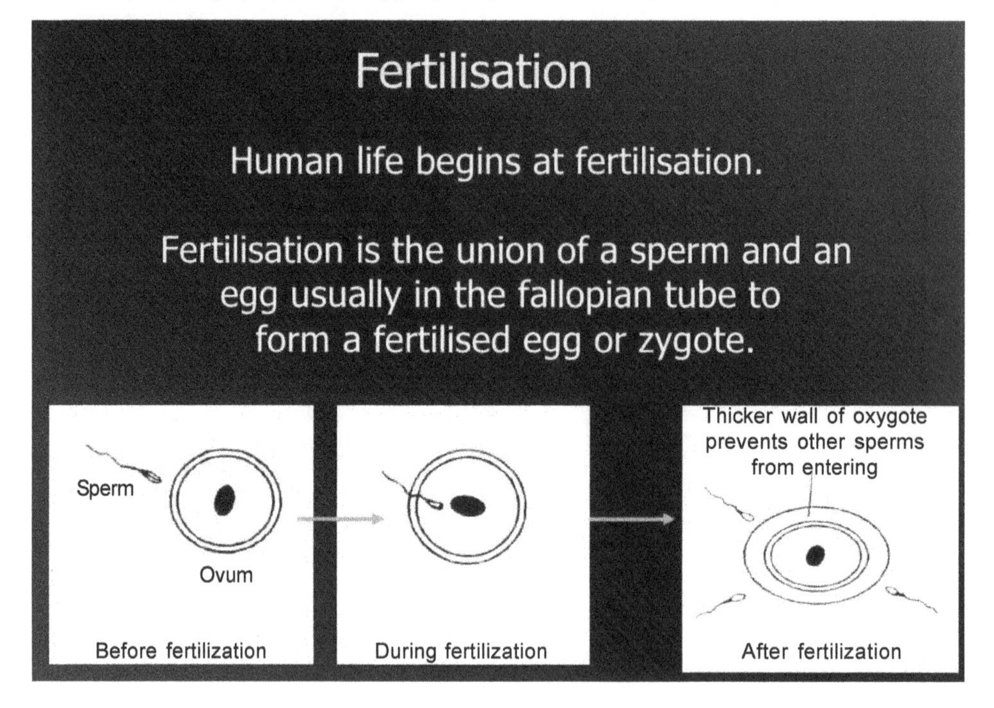

Gestation

- It is also called as a pregnancy
- The condition of female when a developing young is present in the uterus, the condition is called as gestation or pregnancy.
- The period of pregnancy, pregnant period or gestation period is the duration of time which elapses between service and parturition.
- Methods of detecting pregnancy
  1. External signs exhibited by animals
  2. Per-rectal examination
  3. Laboratory tests- hormone detection

Signs of pregnancy

1. Animal not comes in heat after 21 days or concurrent estrus.
2. The animal concerned tends to become sluggish in temperament more tractable.
3. The animal has a tendency to grow fat.
4. Increase in body weight of the animal at the last half of pregnancy due to the development of foetus and hypertrophy of the uterus and mammary gland.
5. Increase in the volume of the abdomen at a later stage of pregnancy.
6. Changes in the mammary gland- As parturition approaches the gland becomes firm, enlarged and glossy, teats take a waxy appearance.
    1. External signs- shining of skin, external signs can be pronounced from 6th month of pregnancy.
    2. Per-rectal examination
        i) CL on one of the ovary.
        ii) Upto 2 months of pregnancy foetus can not felt but slight enlargement of the gravid horn can be detected in heifers.
        At the end of 2nd month of the pregnancy if the free part of the gravid horn is gently pinched through rectal wall, sensation of the foetal membranes slipping between the fingers can be detected.
        iii) In third month the size of the gravid horn is more detectable, can easily be compared with the non-gravid horn.
        iv) In fourth month cotyledons can be felt. At the end of 4th month uterine artries start enlarging. The middle uterine artery if compreesed between the finger & the thumb, continuous vibrating fremitus(uterine thrill) can be detected.
        v) From 5th month of pregnancy, the uterus sinks below the pelvic brim untill the middle of the 5th month.
        vi) From 7th month to 9 or 10th month foetus can be felt.
    3. Laboratory tests
        i) Use of ultrasonic devices and techniques.
        ii) Progesterone assay (in milk & plasma).
        iii) Pattern of vaginal smear- vaginal smear first stain & fixed and observe under the microscope. The more proportion of large nucleated sperical cells is indicating of pregnancy.
        iv) Immunological techniques.
        v) Faetal electrocardiography.
        vi) Barium choride test.

Parturition in cows & buffaloes

- It is the expulsion of the foetus and its membranes from the uterus through the birth canal by natural forces and in such a state of development that the foetus is capable of independent life.
- Process of parturition in cows & buffaloes is called as **calving**.

Signs of parturition

1. Labour pain.
2. Discomfort & general disturbance.
3. Prior to parturition, the pelvic ligament of the cow, especially sacrosciatic becomes more relaxed causing sinking of the croup muscles. Tendons and muscles get relaxed causing depressed or hollow appearance on either side of tail head.
4. The vulva becomes more flaccid (2 to 6 times more).
5. Udder becomes elongated & swollen.Udder appears hard and filled with colostrum.
6. Stringy thick types of mucous from vagina.
7. Teat appears glossy and waxy.
8. Cow likes lonesome place.
9. Uneasiness.
10. Animal sits down and get up frequently.
11. Straining, passes small quantity of faeces frequently.

Causes of parturition

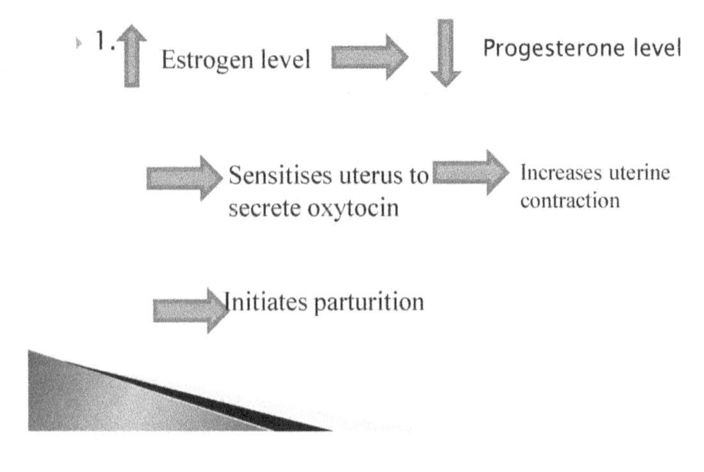

12. Increase in size of foetus leads to fatty degeneration of placental cells which hampers the interchange between foetus and dam.
13. Creates slight irritation in uterus.
14. Leads to contraction of uterus & expulsion of foetus.

Stages of parturition

1. Preliminary stage
2. Dilation of the cervix stage
3. Expulsion of foetus satge
4. Expulsion after the birth

1. Preliminary stage-it is characterised by

- It continues for about hours to days.
- Swelling of udder & clear waxy fluid oozes from teats or may expelled by hand pressure.
- Genital organs become swollen & reddish.
- A clear stringy mucous secreted, which soils tail & hind quarters.
- The quarters droop along with slackening of the **muscles & ligment of pelvic region.**
- **Animal feels uneasy, bellows and become excited.**

2. Dilation of the cervix stage

- marked increase in uneasyness & labour pain.
- Show signs of pain at abdomen.
- Animal may lie & rise again several times.
- Cow becomes anxious, pulse rate increases & distressed breathing & rapid respiration.
- After about 1/2 to 3 hours water bag appears at vulva.
- Forefeet of the foetus will appear in that water bag.
- At this time cervix is fully dilated.

3. Expulsion of foetus satge

- **In this phase co**mpletion of dilation of os-uteri & delivery of the foetus.

- **The back is arched & chest is expanded**
- **The** muscles of abdomen become broad & hard with each labour pain.
- The bag bursts & quantity of fluid is thrown off .
- The calf normally comes with the front hoofs first.
- Normal presentation of foetus: Forelimbs extended and calf head lies between the knees, straight body and hindlimbs. Such a presentation needs no attention.
- Abnormal presentation, position and posture: Any deviation from the normal presentation of calf if occurs, help of expert veterinarian should be taken to relieve dystocia.
- When the hoofs & the nose are at genital of the cow, the head of the cow is at the pelvic which have to pass through the small pelvic opening. This is the moment of the supreme effort & of the greatest point of labour pain.
- At last the uterine contractions combined with the additional abdominal force on the uterus results in the driving the foetus outside.
- **The rectum** forcibly discharges its contents & urinary bladder does likewise
- **At each contraction, the water bag protrudes further & further from** the vulva till the front hoof is visible.

4. Expulsion after the birth

- In uterus the attachment of placenta with uterus is limited to the cotyledons.
- After the expulsion of the calf the uterus tends to throw out the placental membranes which is now merely a foreign body.
- As a result of uterine contractions the placenta separates from the cotyledons & passes into the vagina, where from it expelled.
- Expulsion of placenta within 6 to 8 hours after parturition may be regarded as normal.

  Parturition period: In case of normal presentation and subsequent normal partirition after onset of labour pain- this period is between 2 to 3 hrs (In first calving period somewhat lengthens i.e. 4 to 5 hrs.).

CHAPTER 23

METHODS OF SELECTION OF DAIRY ANIMALS

Selection

Selection is defined as a process in which certain individuals in a population are preferred over others for producing next generation.

Selection is the tool in the hand of the breeder to improve the performance of the animal.

- Selection does not create new genes, but it only increase frequency of desirable genes.
- Selection may be natural or artificial.
- Natural selection goes by time with nature.
- Man, aimed at improving genetic potential of farm animals, controls artificial selection.

Methods of Selection

1. Individual selection
2. Pedigree selection
3. Family selection
4. Progeny testing
5. Unified score card system
6. Body Condition Score (BCS) system

1. Individual selection

- Selection on the basis of individual phenotypic performance is called individual selection.
- It is the most commonly used basis for improvement in livestock.
- Characters like body type, growth rate are evaluated directly from the individual animal performance.

Limitations of Individual selection

1) Some important traits like milk production, maternal abilities in cows are expressed only in females.
2) The performance records for milk and other maternal qualities are available only after sexual maturity is reached.
3) When the heritability of a trait is high, individual merit is a poor indicator of breeding value.

2 Pedigree selection

- Selection on the basis of performance of the ancestors is called as pedigree selection.
- Pedigree selection is very useful when the traits selected are highly heritable.
- If a performance record of individual is available, the addition of pedigree information usually adds little to accuracy of estimates of breeding value of individual.
- Pedigree selection is especially useful for early selection of individuals as in case of selection of young bulls for progeny testing.
- **Limitations of pedigree selection**
- The environment under which ancestor records were made several years ago are quite different from the existing environmental conditions when an individual is evaluated for selection.

3. Family selection

- When individual's performance is also included in calculating the sibs average performance, it is called family selection.
- Family represents a group of animals having common genetic relationship.
- Generally full sibs and half sibs are the most common collateral relatives, whose records are often used to estimate the breeding value.
- Family selection is very useful in case of traits with low heritability.

4. Progeny testing

- Selection of the individuals on the basis of average performance of their progeny is called progeny testing.
- It is the estimation of an individual by evaluating its off springs.
- It is very useful tool in evaluating breeding worth of dairy cattle.
- It offers best means of achieving genetic improvement in traits of moderate to low heritability.

Advantages

1. It offers best means of achieving genetic improvement in traits of moderate to low heritability.
2. The rate of progress achieved by this method is double to that possible by phenotypic selection.
3. Progeny testing is generally used for selecting males as a large number of progeny can be obtained for each male, while the number of progeny produced by a female is limited.

Limitations of Progeny testing

1. Progeny testing requires prolonged generation interval.
2. It is also very expensive since a large number of animals are to be performance recorded.
3. For practical genetic consideration, the number of unselected daughters studied to evaluate a bull should be between 30 and 50.

5. Unified score card system

- The unified score card system gives a good index of Dairy confirmation of animals

A.	**General appearance : 18**	
1	Size of animal- ideal to breed and age	3
2	Form of animal- symmetrical or stylish	2
3	Dairy character –lean ,angular	5
4	Skin quality-thin, pliable, hairs smooth, fine and soft	4
5	Temperament- active vigorous disposition and docile	4

B.	**Head and neck : 09**	
6	Muzzle- wide nostrils and large	2
7	Face – clean cut, facial veins prominent	1
8	Forehead- wide fine at poll	1
9	Horns- fine, typical of breed	1
10	Neck-Slender, medium length	1
11	Eyes- Large, bright prominent	1
12	Ears- Typical of breed size, well set	1
13	Dewlap- Thin, light, graceful folds	1
C.	**Fore-Quarters: 07**	
14	Withers-Clean refined, free from fleshiness	3
15	Shoulders- light oblique, well attached free from fleshiness	2
16	Legs-Straight, well apart, fine and smooth	2
D.	**Body :20**	
17	Chest, wide, deep fore-flank full	6
18	Back-Straight, strong, vertebra, well defined	4
19	Lion- Broad strong, leveled, free from flesh	3
20	Ribs – wide apart and well sprung	6
21	Flanks- Thin, deep and full	1
E.	**Hind Quarters :12**	
22	Hip bones- Prominent, and wide apart	2
23	Rump-Long, wide leveled.	3
24	Pin bones-Prominent, and wide apart	2
25	Tail setting –Long fine, tapering	1
26	Thigh- Thin, widely separated and incurring	2
27	Hind legs- Straight carried well apart, fine shank	2
F.	**Mammary Development: 34**	
28	Udder, (a) Shape	
	(i) Fore Udder: Full attached forward	5
	(ii) Rear Udder: Full Attached, high and wide	5
	(B) Symmetry: Quarters even balanced, floor of udder leveled	3
	(C) Capacity: Large, Texture, pliable, Free from fat and fibrous tissue	12
29	Teats: medium sized, squarely placed	4
30	Milk Veins: Long, Tortuous	3
31	Milk Wells: Large, numerous	2
	Total	**100**

- Classify the cows based on Dairy type as follows

Grade Score

- Excellent 90 and Above
- Very Good 85 to 90
- Good 80 to 85
- Acceptable 70 to 80
- Fair 60 to 70
- Poor below 60
- Animals scoring below 75 must be discarded.

6. Body Condition Score (BCS) system

- Body condition is defined as the ratio of the amount of fat to the amount of non-fatty matter in the body of the living animal.
- Body conditions are reflection of the fat reserves carried by the animal.
- The ability to estimate the body condition more accurately and relate it to milk and milk components production would help the farmers in the selection of dairy animals and to increase the overall efficiency of feeding and management of dairy animals.
- It is helpful as a cheap tool for the selection of dairy animals.
- BCS was observed to be highly correlated with both body weight and heart girth.
- Cows with BCS above 3.5 had more heart girth which is a true indicative of health condition and productive performance.
- Cows with BCS of below 2 will be considered as under conditioned and above 4.5 as over conditioned

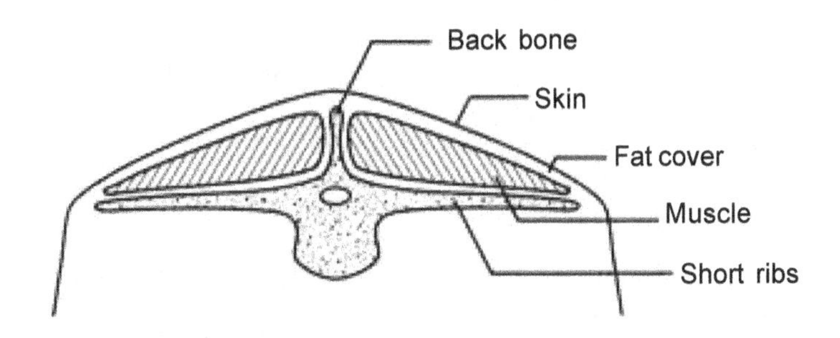

Condition Score 1

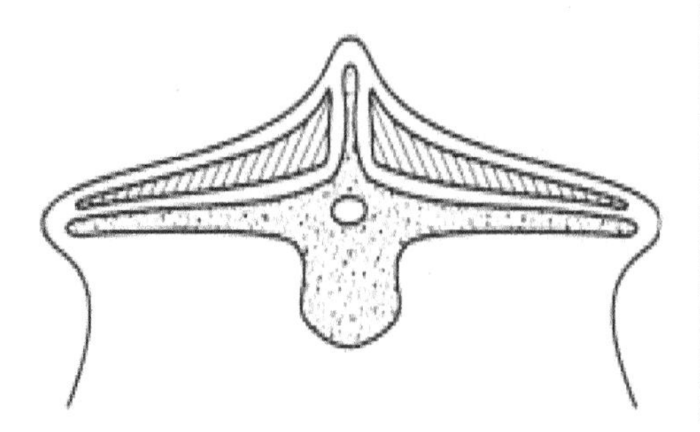

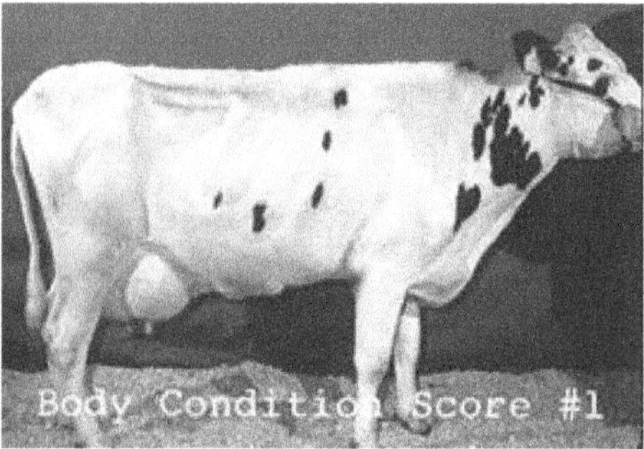

This cow is emaciated. The ends of the short ribs are sharp to the touch and together give a prominent shelf-like appearance to the loin. The individual vertebrae (spinous processes) of the backbone are prominent. The hook and pin bones are sharply defined. The thurl region and thighs are sunken and in-curving. The anal area has receded and the vulva appears prominent.

Condition Score 2

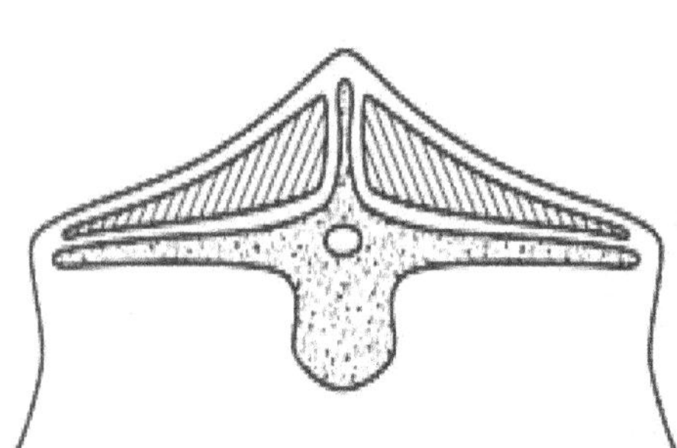

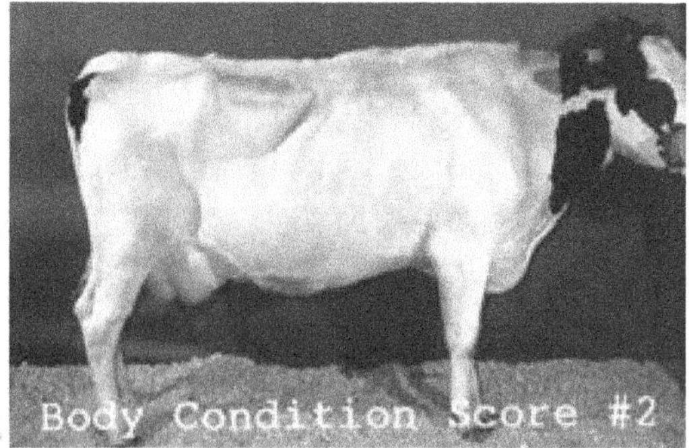

This cow is thin. The ends of the short ribs can be felt but they and the individual vertebrae are less visibly prominent. The short ribs do not form as obvious an overhang or shelf effect. The hook and pin bones are prominent but the depression of the thurl region between them is less severe. The area around the anus is less sunken and the vulva less prominent.

Condition Score 3

A cow in average body condition. The short ribs can be felt by applying slight pressure. The overhanging shelf like appearance of these bones is gone. The backbone is a rounded ridge and hook and pin bones are round and smoothed over. The anal area is filled out but there is no evidence of fat deposit.

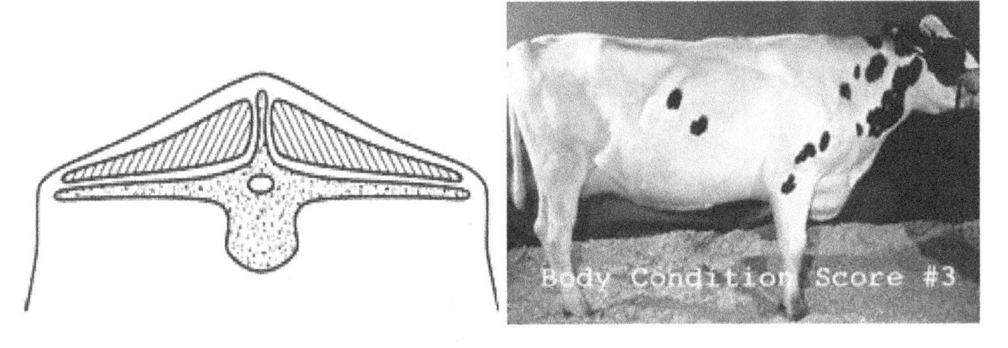

Condition Score 4

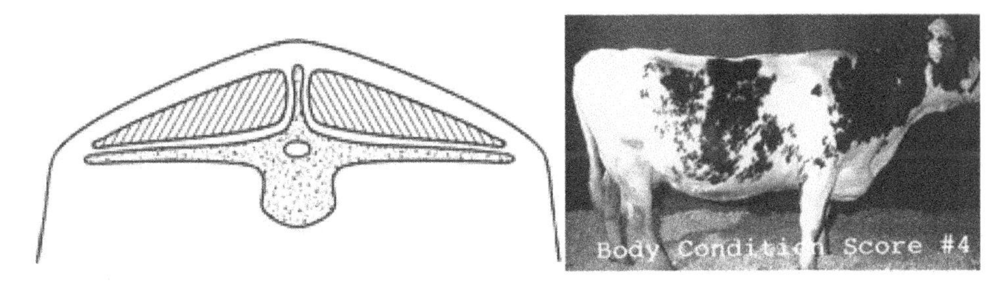

A cow in heavy condition. Individual short ribs can be felt only when firm pressure is applied. Together they are rounded over with no shelf effect. The ridge of the backbone is flattening over the loin and rump areas and rounded over the chine. The hook bones are smoothed over and the span between the hook bones over the backbone is flat. Area around the pin bones is beginning to show patches of fat deposit.

Condition Score 5

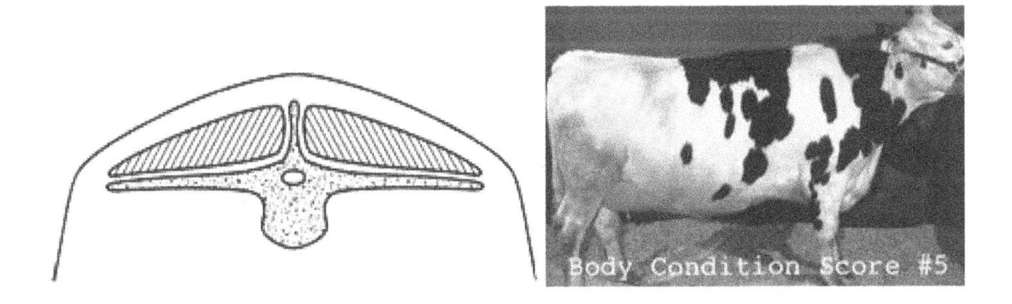

A fat cow. The bone structure of the topline, hook and pin bones and the short ribs is not visible. Fat deposits around the tailbone and over the ribs are obvious. The thighs curve out, the brisket and flanks are heavy and the chine very round.

Table 2. Suggested Body Condition Scores for Cows by Stage of Lactation

Stage of lactation	DIM	BCS Goal	BCS Min.	BCS Max.
Calving	0	3.50	3.25	3.75
Early Lactation	1 to 30	3.00	2.75	3.25
Peak Milk	31 to 100	2.75	2.50	3.00
Mid Lactation	101 to 200	3.00	2.75	3.25
Late Lactation	201 to 300	3.25	3.00	3.75
Dry Off	> 300	3.50	3.25	3.75
Dry	- 60 to -1	3.50	3.25	3.75

CHAPTER 24

STRUCTURE AND FUNCTION OF MAMMARY SYSTEM

Structure of mammary system

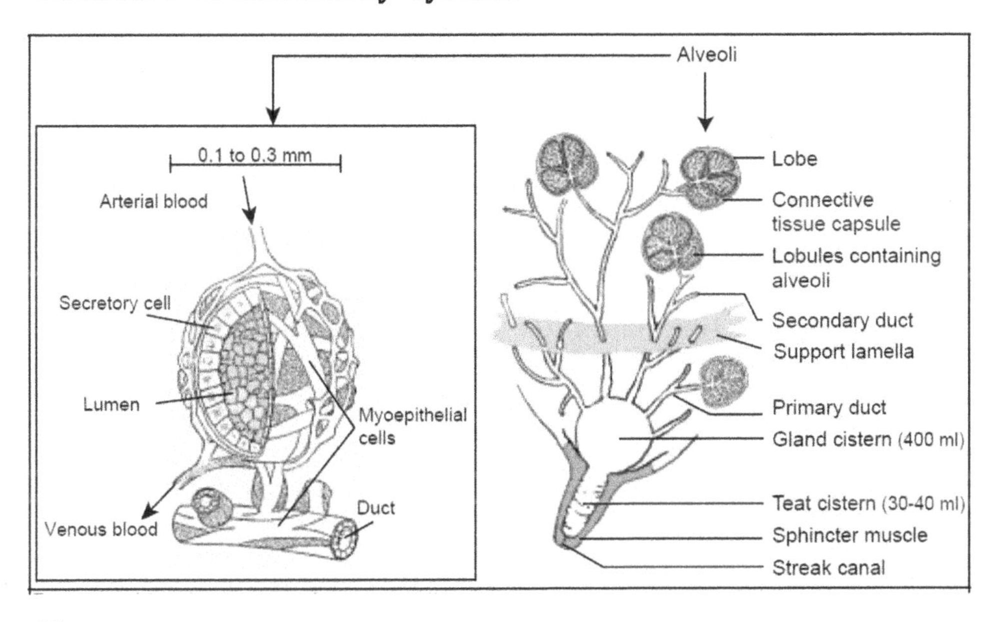

Mammary system

- The udder of a cow is an organ designed to produce and offer a newborn calf easy access to its mother's milk. It is suspended outside the wall of the rear abdomen and thus it is not restrained, supported, or protected by any skeletal structures.
- The udder of a cow is made up of four mammary glands or "quarters."

- Each quarter is a functioning entity of its own which operates independently and delivers the milk through its own teat.
- Generally, the rear quarters are slightly more developed and produce more milk (60%) than the front quarters (40%).

Structure of mammary system

- **Support system.** A set of ligaments and connective tissue maintain the udder close to the body wall.
- Strong ligaments are desirable because they help to prevent the occurrence of pendulous udder, minimize the risk of injuries, and avoid difficulties when using milking equipment.

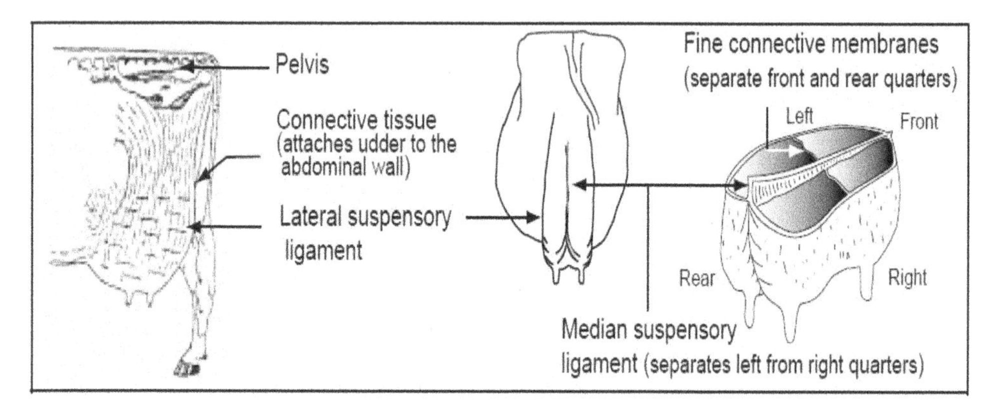

Secretory and duct system. The udder is known as an exocrine gland because milk is synthesized in specialized cells grouped in alveoli and then is excreted outside the body through a duct system that functions like the tributaries of a river.

Blood supply and capillary structures. Milk production demands a lot of nutrients that are brought to the udder by the blood.

- To produce 1 kg of milk, 400 to 500 kg of blood must pass through the udder.
- In addition, the blood carries hormones that control udder development, milk synthesis, and the regeneration of the secretory cells between lactations (during the dry period).

Lymph system. Lymph is a clear fluid that comes from tissues highly irrigated by blood.

- The lymph helps to balance the fluid flowing in and out of the udder and helps to combat infections.

- Sometimes the increased blood flow at the onset of lactation leads to an accumulation of fluid in the udder until the lymph system is able to remove the extra fluid. This condition, referred to as udder edema, is more prevalent in first-calf heifers and older cows with pendulous udders.

Innervation of the udder. Nerve receptors on the surface of the udder are sensitive to touch and temperature.

- During the preparation of the udder for milking, these nerves are triggered and initiate the "milk let down" reflex that allows the release of milk.
- Hormones and the nervous system are also involved in the regulation of blood flow to the udder. For example, when a cow is started or feels physical pain, the concerted action of adrenaline and the nervous system decreases blood flow to the udder, inhibits the "milk let down" reflex and lowers milk production.

CHAPTER 25

MILK SECRETION AND MILK LET DOWN

Milk secretion

- Milk secretion by the secretory cells is a continuous process that involves many intricate biochemical reactions.
- During milking, the rate of milk secretion is somewhat depressed, but it never stops completely.
- Between milkings, the accumulation of milk increases the pressure in the alveoli and slows down the rate of milk synthesis.
- As a result, it is recommended that high-producing cows be milked as close as possible to 12 hour intervals (the highest milkers should be milked first in the morning and last in the evening). More frequent ejection of milk reduces the pressure build-up in the udder, and for this reason milking three times a day can increase milk yield by 10 to 15%.
- **The use of glucose by a secretory cell.** Although glucose in the diet is entirely fermented in the rumen in volatile fatty acids (acetic, propionic and butyric acids), it is needed in large amounts by the lactating udder. The liver transforms propionic acid back into glucose which is transported by the blood to the udder where it is taken up by the secretory cells. Glucose can be used as a source of energy to the cells, as the building block of galactose and subsequently lactose, or as the source of the glycerol needed for the synthesis of fat.
- **Synthesis of lactose.**
- The synthesis of lactose is controlled by a two-unit enzyme called lactose synthetase. The sub-unit á-Lactalbumin is found in the milk as a whey protein.

- **Regulation of milk volume.** The amount of milk produced is controlled primarily by the amount of lactose synthesized by the udder. Lactose secretion into the cavity of an alveolus increases the concentration of dissolved substances (osmotic pressure) relative to the other side of the secretory cells where the blood flows. As a result, the concentration of dissolved substances on each side of the secretory cells is balanced by drawing water from the blood and mixing with the other milk components found in the cavity of the alveolus. For normal milk, a balance is reached when there is 4.5 to 5% lactose in the milk. Thus lactose production acts as a "valve" that regulates the amount of water drawn into the alveoli and therefore the volume of milk produced.
- The effect of the diet on milk production may be easily seen:
    1) The amount of energy (i.e., concentrates) in the diet influences propionate production in the rumen.
    2) The propionate available influences the amount of glucose synthesized by the liver.
    3) The glucose available influences the amount of lactose synthesized in the mammary gland.
    4) The lactose available influences the amount of milk produced per day.
- **Synthesis of protein.** The caseins found in the milk are synthesized from the amino acids taken up from the blood under the control of the genetic material (DNA). These proteins are packed in micelles before they are released in the lumen of the alveolus. Genetic control of milk synthesis in the alveoli comes from the amount of α-Lacto-albumin synthesized by the secretory cells.
- **Synthesis of fat.** Acetate and butyrate produced in the rumen are used, in part, as the building blocks of the short-chain fatty acids found in milk.
- The glycerol needed to "unite" three fatty acids into a triglyceride comes from glucose. About 17-45% of the fat in the milk is built from acetate and 8- 25% from butyrate.
- Diet composition has a strong influence on milk fat concentration. Lack of fiber depresses the formation of acetate in the rumen, which in turn results in the production of milk with a depressed concentration of fat (2-2.5%).
- Lipids mobilized from body reserves in early lactation are another building block for milk fat synthesis.
- However, in general, only half the amount of fatty acids in milk fat is synthesized in the udder, the other half comes from the predominantly long chain fatty acids found in the diet. Thus milk fat composition may be altered by manipulating the type of fat in the cow's diet.

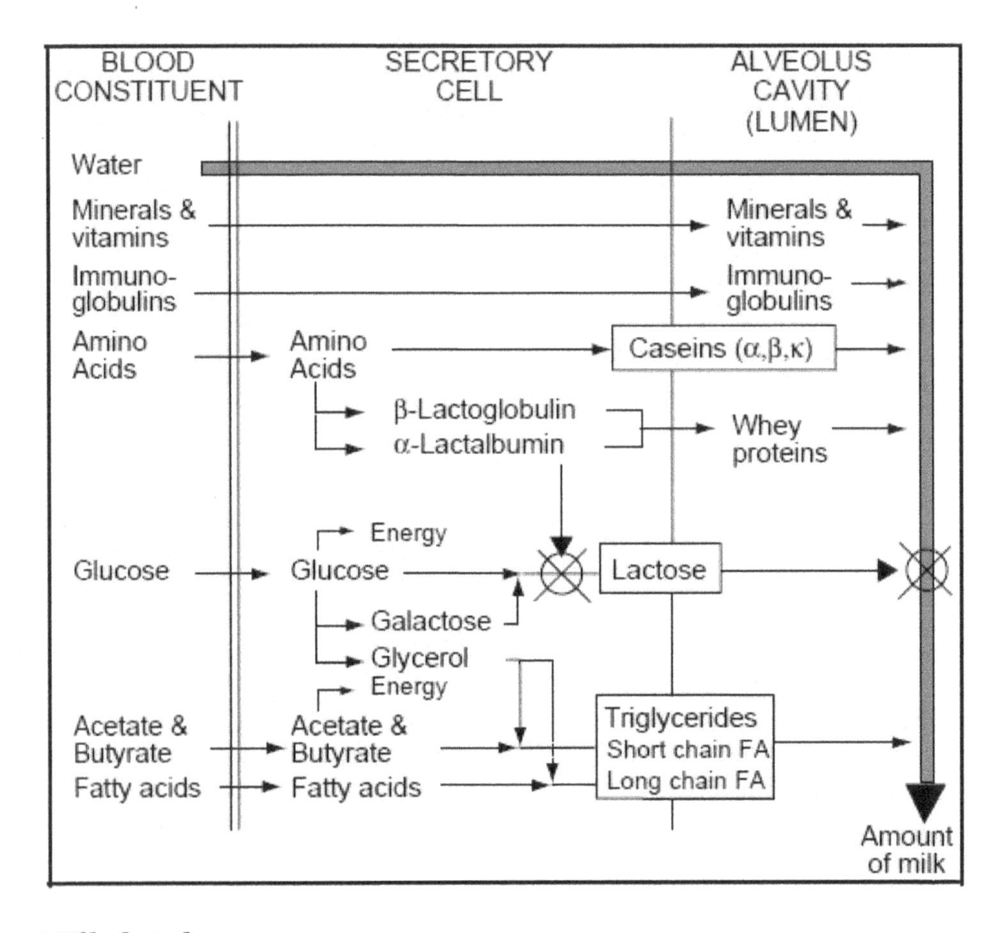

Milk let down

- As the secretion of milk is continuous process, most of the milk secured at any milking is already present in the udder when milking starts. A small amount in the teat and gland cisterns, but most of it is in the alveoli and ducts to the gland cisterns, where it removed by either hand milking or machine milking.
- Oxytocin is secreted from the posterior pituitary gland and is a must for the ejection of milk or milk let down.
- Stimulation of udder by washing or suckling » Brain activates pituitary gland » Pituitary gland secretes oxytocin into blood stream » Oxytocin causes milk let down.

CHAPTER 26

METHODS OF MILKING, PROCEDURE OF MILKING AND PRACTICES FOR QUALITY MILK PRODUCTION

Milking

- Milking is an act of removing milk from the udder
- Oxytocin hormone is required for letting down of milk

Milking methods

- Two methods of milking
    1. Hand milking
    2. Machine milking

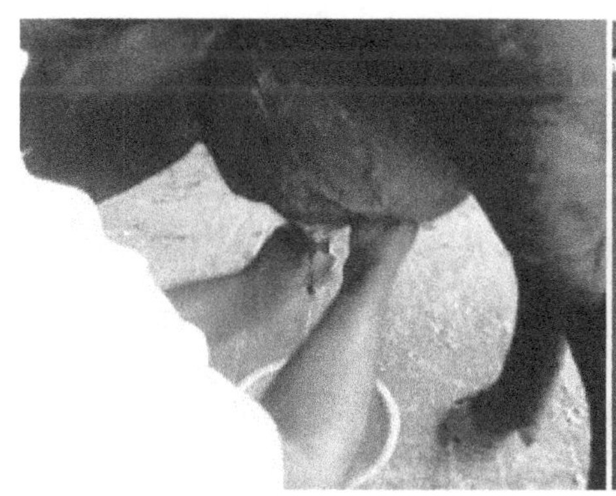

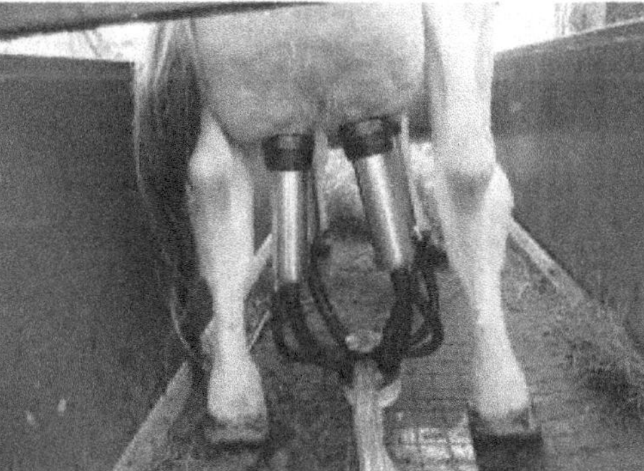

Hand milking

1. **Hand milking by Stripping:** Stripping is done by firmly holding the teat between the thumb and the fore-finger and drawing it down the length of the teat and the same time pressing it to cause the milk flow down in a stream.

MILKING METHOD

1 2 3 4 5 6 7 8 9 10

Step 1. Hand washing
Step 2. Udder washing and wiping
Step 3. Teat massaging
Step 4 to 9. Full hand milking actions
Step 10. Stripping

2. **Hand milking by fisting or full hand milking:** It is done by grasping the teat with all the five fingers and pressing it against the palm.

   Fisting method removes milk faster than the stripping because of no loss of time in changing the position of the hand

   As full hand milking stimulates the natural suckling process by calf, it is superior than stripping.

3. **Knuckling Method of hand milking:** In this method the milk is removed by bending thumb against teat and pressing.

   Milking by Knuckling method injures the teat tissue. Therefore avoid milking by this method

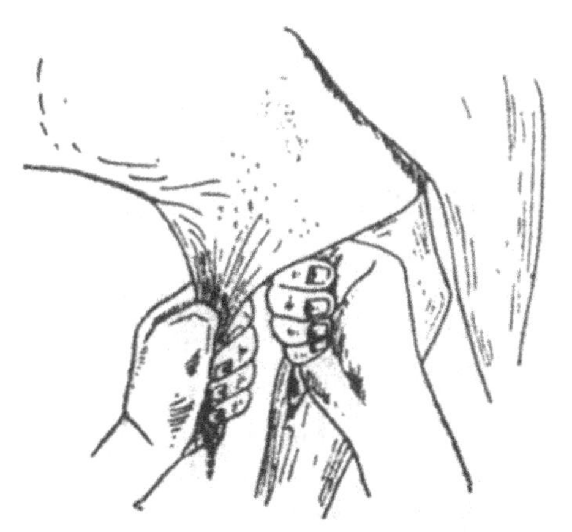

Knuckling Method of hand milking

Machine milking

There are 3 types of machines present in market

1. Electric operated milking machine
2. Hand operated milking machine &
3. Three in one milking machine

Basic functions of milking machine

- 1. It opens the teat canal through the use of partial vacuum, allowing the milk to flow out of the teat cistern through a line to a receiving container.
- 2. It massages the teat, which prevents congestion of blood and lymph in the teat.

Precautions during machine milking

- Teat cups should not be kept for long time.
- Machine should be operated with optimum vaccum.

Procedure of Milking

1. Clean the milking pen.
2. Wash the udder with Potassium permagnate solution or weak antiseptic solution.
3. Dry udder with help of clean soft cloth or towel.
4. Wear clean cloths.

5. Wear cap on head to avoid entry of hairs in milk.
6. Clean the hands with dettol or savlon.
7. Remove the first strips of milk & check blockage or abnormality in milk.
8. Remove the milk by hand milking or machine milking.

   In case of machine milking teat cups are placed on teats. Then the machine is operated at 10-15 inches vaccum with a pulsation rate of 50/ min. The negative pressure created by machine evacuates the milk from udder. Teat cups are removed as soon as milk ceases to flow.
9. At the end of machine milking remove last strips of milk from udder with stripping. Last strips are very rich in fat.
10. Apply dip cup to avoid entry of pathogens through teat sphincters.

Care to be taken while milking

1. Milking should be gentle, quick and complete.
2. Letting down of milk can be stimulated by gentle massage.
3. Milking should be completed within 7 minutes.
4. Avoid excitement of cow before and during milking.
5. For safe milking secure hind legs with anti-cow kicker or '8'knot.
6. Milking is usually done from left side.
7. The milking of first few strips (2-3) in separate vessel and other in separate container. The first drawn milk strips are collected in strip cup to check any abnormality.
8. Milking time and interval between milkings should be kept constant.
9. Animal should not be disturbed or excited during milking as it may cause inhibition of milk ejection due to action of adrenaline hormone.

Practices for quality/clean milk production

Clean milk' refers to raw **milk** obtained from healthy animals, that has been **produced** and handled under hygienic conditions, it should contain very small number of harmless bacteria, it should be free from hazardous chemical residues.

Reasons to learn and implement clean milk production principles

- Consumers will reject dirty milk at the market. Since you had spent money in feeds, medication and labor, there is no way you will get your money back. Consumers may even learn to boycott your products if the quality issues persist.

- Dirty milk is extremely difficult to handle because it goes bad pretty fast. You will most probably throw it away due to its short shelf life.
- Dirty milk is a rich medium for the transmission of food borne and zoonotic diseases. Cross contamination with such conditions such as mastitis will reduce productivity at the farm.
- Clean milk helps in controlling the spread of infectious diseases like Tuberculosis and Diphtheria etc. You and I know that a sick nation cannot grow.
- The government will lose revenue from the dairy sub-sector. Considering that the dairy sub-sector contributes 30 percent of the national economy, which contributes 10 percent of the Gross Domestic Product (GDP), this will be a substantial loss.
- Producing clean milk will make you a more satisfied farmer because you know that your consumers are getting only the best products.

1. **Maintain herd healthy:** The herd should be free from pathogens that might be spread to human beings through milk e.g. Tuberculosis

   The animals should periodically be checked for contagious diseases (every year).

   Follow the following mentioned instructions to keep animal healthy.

   - Examine your animals regularly to ensure your animals are free from contagious diseases such as brucellosis, mastitis, TB, etc
   - Identify and separate sick animals from the healthy ones. Treat them separately to avoid spreading the disease to the healthy animals.
   - The animals suffering from contagious disease must be kept separate from healthy herd. Sanitary precaution to prevent and control the disease should be adopted.
   - Avoid mixing milk from cows with mastitis with that from healthy cows. Always milk the sick cows last and use their milk to feed the calves.
   - Vaccinate the animals as regularly as is needful for such diseases as foot and mouth (FMD) and Anthrax. Consult your vet for a personalized advice.
   - When cleaning the udders, use disposable towels to dry up the udders. Do not share the towel among many cows.
   - The milker should be in perfect health to avoid contaminating the dairy products.

2. **Cleaning of animals prior to milking:** To prevent dirt from getting into milk Cattle shall be bathed and if bath is not possible dry cleaning by broom /duster should be done

3. **Proper housing for clean milk production**
    - The cow shed should be well constructed to ward off extreme weather, and well-drained to avoid stagnation of water. Stagnation encourages breeding of pathogens, which cause contamination. Always keep the shed clean and dry.
    - Remove all the remains of fodder and feeds. Clean the feeding and watering troughs before every feeding session. Design the cow shed to facilitate removal of wastes. Dispose the manure in a pit or biogas plant.
    - Overlay the windows with a wire mesh and a net to keep off flies.
    - Arrange the farm in a manner to facilitate easy access of the feed store. Avoid pitting up chicken house within the premise of your dairy project, it will contaminate the milk.
    - Ensure each animal has enough space for feeding, sleeping and exercising. Provide sleeping mattresses for the cows to avoid injuries from the hard floors.
    - Remove accumulated cow dung.
4. **Cleaning of surroundings**
5. **Fly control-** flies carry typhoid, dysentry and other contagious diseases.
6. **Cleanliness of milker-** The milker should be free from communicable diseases and must be of clean habbits. Nails must be trimmed, wear clean cloths, do not spit around etc. Milker should wash his/her hands with soap to make them clean and germ free. Milker should avoid contact between milk and his body parts, clothes and other belongings. Chewing and spitting with tobaco, smoking and gutka should be avoided during milking. Sneezing/ coughing towards udder/vessel during milking should be avoided.Milker should not be suffering from any respiratory ailment or contagious disease. Milker should not have any open sores or cuts.
7. **Clean utensils-** The utensils should be free from pathogens. Utensils must be cleaned with warm water & detergents immediately after milking.
    - Clean and sanitize all the milking/milk handling vessels before and after every milking session. Usually, after you have effectively cleaned the equipment after milking, you will only need to sanitize it before milking. Sanitize the equipment at least 20 minutes before you start milking. Do not forget to pour out the sanitizer before adding milk into the container.
    - Use detergents that are easy to rinse off the equipment. Soap is discouraged because it is difficult to completely rinse. You can use sodium hydroxide as a sanitizer.

- After cleaning, invert the equipment/utensils to dry up; that way, few bacteria settle in it.
- Use separate vessel for washing of udder and teats & for milking.

8. Type of milk pan- Sanitary milking pails with dome shaped top should be used for milking.
9. Straining- Straining is done to remove sediment and other foreign materials. It should not be used as a cover for unclean milk. If cloth is used, it should be washed and dried daily otherwise dirty cloth will spoil the quality of milk further rather than to improve it.
10. Cool and store milk properly- After milking, milk should preferably be cooled by keeping milking pails in cold water in winter and water added with ice cubes in summer.
    - Cool the milk to less than 5°C immediately after milking. Most bacteria will not multiply at this temperature. Be careful to ensure you cool the milk immediately after milking because if you wait, it will develop acidity, which you cannot remove by cooling.
    - Deliver the milk to the dairy within two hours after milking if you do not have the cooling facilities.
11. **Feeding of animals:** Provide balanced feed to the animal. Provide dust free concentrate at the time of milking to make animal busy in eating and remain quite during milking

    In any case, always observe the following points when managing a dairy animal for clean milk production:

    - Ensure the feeds and fodder contains no pesticides, insecticides, fungicides, herbicides, aflatoxins, and metals.
    - Always store your animal feeds with the minimum allowable moisture content.
    - Handle farm chemicals with a lot of care because they can easily find their way into the milk
    - Avoid giving the animal silage and wet crop residues during milking to avoid imparting any off-flavor to the milk.
    - Ensure the animal gets adequate supply of key minerals and vitamins
12. **Awareness and training:** Educational aids and programmes should be organized for the farmers for making them aware of the importance of clean milk production. This should be in the form of charts/posters displayed at village, society and milk collection centers. Make them aware of the correct handling of the milk from udder to reception dock, maintenance of hygienic environment, clean utensils to availability of milk cooling bulk tanks and coolers.

13. In case animal is under treatment, discard the milk during the withdrawal period of the treatment. Do not bring the milk to DCS/MPI, if the cattle is suffering from any disease.
    - Milking machines are capable of milking quickly and efficiently without injuring the udder
    - Mostly time required for machine milking is 7 minutes.

Difference between full hand milking and stripping

Sr.No.	Full hand milking	Stripping
1	In this method, teat is held in fist or four fingers around teat.	In this method teat is held between thumb and finger.
2	By this method Milking is quick	By this method milking is slow.
3	It is similar to natural suckling process of calf	It is different than the natural suckling by calf
4	It causes less irritation to teats hence animal is at comfort during milking.	It causes more irritation to teats, hence animal is less comfortable at milking.
5	Suitable for high yielders	Suitable for very low yielders.
6	Suitable for animal having long teats	Suitable for animal having small teats
7	It is generally preferred for whole milking of animals	It is generally preferred to remove last few strips of milk after milking
8	More power is required for milking	Comparatively less power is required for milking

Difference between Hand milking and machine milking

Sr.No.	Hand milking	Machine milking
1	Milking of animals can be carried out by hands	Milking is carried out by milking machine
2	It is less expensive method	It is more expensive method
3	Milking is slow	Milking is rapid
4	Labour requirement is more	Labour requirement is less
5	It is suitable for small size of herds	It is suitable for bigger herds
6	Milk production is less hygienic	Hygienic milk production is possible
7	More time is required for milking per animal	Comparatively less time required for milking per animal

[Table Contd.

Contd. Table]

Sr.No.	Full hand milking	Stripping
8	It is possible in presence or absence of electricity, diesel engine	Machine milking requires electricity or fuel to run the milking machine
9	Painful milking if knuckling method is used for milking	No any pain during milking with machine
10	More chances of udder infection	Less chances of udder infection

Drying off cows

It is most essential practice in case of high yielder animals. Dry period of 60-90 days is required as a rest period to udder.

Thin cows should have longer dry periods than those carrying more flesh. Shorter dry period observes in case of low yielders.

Reasons for drying the animal

1. To rest the organs of milk secretion.
2. To permit the nutrients in feed to be used in developing the foetus instead of producing milk.
3. To enable the cow to replenish in her body the stores of minerals which have become depleted through milk production.
4. To permit her to build up a reserve of body flesh before calving.

Methods of drying off

There are three methods of drying off of cow

1. Incomplete milking
2. Intermittent milking
3. Complete cessation

1. **Incomplete milking:** In this method, do not draw all the milk from udder at milking time does the first few days after the drying off period has begun. Later they milk intermittently but never completely. After the production decreases to only a few liters daily, milking is stopped.
2. **Intermittent milking:** In this method the milking of animal carried out once a day for a while, then once in every next day and finally milking will be

stopped altogether. Milking of animals only one times a day by gap of one day.

3. **Complete cessation:** This method is recommended for a cow which produces 10 lit. of milk/day. By this method, the udder fills until pressure increases enough to stop secretion inside the udder. After the cessation of secretion of milk is gradually reabsorbed from the gland until it becomes completely dry. The cow should not be milked during the stage of re-absorption as this releases the pressure within the gland and secretion is again initiated, resulting in a prolonged period of drying off.

CHAPTER 27

FACTORS AFFECTING MILK COMPOSITION OF ANIMALS

The milk composition of different animals is differs in percentage of nutrient content. Milk from all animal contains the same kind of constituents. Milk from individual cows shows greater variation than mixed herd milk. The variation is always greater in small herds than in large ones.

Normal milk composition of cow, buffalo and goat milk

Sr.No.	Constituents	Buffalo milk	Cow milk	Goat milk
1	Water%	82	87.5	87
2	Fat%	7.9	3.7	4.0
3	Protein%	4.5	3.3	3.6
4	Lactose %	4.8	4.7	4.5
5	Minerals %	0.80	0.80	0.80
6	Energy, kcal	117	67	72
7	Calcium(mg)	210	120	170
8	Phosphorus (mg)	130	90	120
9	Iron (mg)	0.20	0.20	0.30

Factors affecting the composition of milk

1. Species of animal
2. Breed
3. Individuality
4. Interval of milking

5. Completeness of milking
6. Frequency of milking
7. Irregularity of milking
8. Day to day milking
9. Disease and abnormal condition
10. Portion of milking
11. Stage of lactation
12. Milk yield
13. Feeding of animals
14. Season
15. Age of animal
16. Condition of cow at calving
17. Excitement
18. Adulteration of drugs and hormones

1. **Species of animal-** Different species of animal have different milk composition according to their genetic makeup and digestive physiology of that animal.
2. **Breed:** - In general breed produces more amount of milk contains low fat. And lower milk producing breeds have higher fat in milk.

e.g. In average Jersey milk the fat constitutes 35% of the total solids, while in HF milk fat represents only 28 % of the total solids.

The percentage of ash in the total solids of HF milk is greatest of all i.e. Gurency, Jersey, Ayreshire

3. **Individuality**: - Each individual cow produce milk of its characteristic composition which is differs from other cow within the same breed and same age.
4. **Interval of milking**:- Longer the interval of milking greater the quantity of milk and lower the fat% and vice-versa. When cow is milked at 12 hrs. Interval, little variation is found as some animal have higher fat% in the morning milk. Fat% variation is about 0.75 to 1.00% in 14 to 15 milkings. In short, longer the period of milking fat% will be less.
5. **Completeness of milking: -** If the cow is completely milked, fat percentage is more in milk and vice-versa. First milk drawn from cow is low in fat while last milk generally called stripping is very high in fat.

6. **Frequency of milking:** - If the cow is milked more frequently than the normal (2 time milking) milk production is slightly increase but there is no great effect on the fat percentage.
7. **Irregularity of milking:** - Frequent changes in the time and interval of milking, there is decrease in fat percentage and milk production.
8. **Day to day milking:** - It shows variation for the individual cow only (milch animal should be milked daily at a fix time and interval otherwise).
9. **Disease and abnormal condition:**- These tend to alter the composition of milk and also the milk product. If the cow is suffering from mastitis, fall in milk yield and salt percentage in milk increases.
10. **Portion of milking:**- Fore milk is lowering in fat contain(less%) while last stripping contain more fat percentage (close to 10%). The other milk constituents only slightly affected (on fat free basis).
11. **Stage of lactation:**- Composition of colostrums differ from normal milk in that it is lower in water, sugar and fat and higher in casein, albumin, globulin and ash.

At the end of lactation fat globule size decrease and salt percentage in milk increases. Milk has an objectionable taste and odour after standing 12 to 20 hrs. After milking, this will happen due to action of lipase enzyme on fat globule; this change is described as bitter or rancid.

12. **Milk Yield:** - For a single cow, if milk production increases fat% decreases and vice-versa.
13. **Feeding of animals:**- If milking animal is provided with good quality concentrate and green fodder; the milk production as well as fat percentage increases and vice-versa. Feeding of only green fodder increases milk production but decreases fat % in milk. Feeding of dry roughages leads to increases milk fat percentage. Colour of milk is affected by type of feed given to animal. Volatile oil from onion, turnips, rape and other plant which convey characteristics taste of milk.
14. **Season:** - The percentage of both fat and SNF shows variation during different season especially in winter (Dec.-January) season. Milk production as well as fat percentage increases and in summer season milk production as well as fat percentage decreases.
15. **Age of animals:** - Fat percentage decreases as cow grows older. The fat% increases slightly in 1st and 3rd lactation period observed in high yielding milking animal. As the age advances, after 11th lactation the fat% decreases. As 15 yrs. of age fat percentage in decreases.

16. **Condition of cow at calving:-** If the cow is in good physical condition during calving, milk yield of high fat percentage than the cow is in poor physical condition during calving.

    Fattiness of cow at the time of calving gives temporary effect on increase in size of fat globule is gives 1-2 times more fat than normal (up to 15 days).

17. **Excitement**: - Both yield and composition of milk is affected during the period of excitement.

18. **Administration of drugs and hormones**: - Certain drugs may affect temporary change in the fat percentage, injection or feeding of hormones result in increase of both milk yield and fat percentage.

CHAPTER 28

CLEANING AND SANITATION OF MILKING EQUIPMENTS

Cleaning or washing of utensils

Cleaning or washing of dairy utensils and equipments can be defined as removing of the soil (dirt) from contact surfaces of dairy equipments.

Cleaning dairy utensils include following steps-

- Utensils cleaning using ordinary tap water.
- Scrubbing with 2% soda solution [temp.60^0 Celsius] of dairy detergent.
- Utensils washing using hot water.
- Sterilization [steam of 200 PPM Chlorine solution].

Desired characteristics of washing solution-

1] Complete solubility in water.

2] Non toxic to hands, skin, utensils, equipments.

3] Should have good penetrability.

4] Should have germicidal property for killing harmful microbes.

5] Should have good buffering capacity to maintain pH.

6] Good wetting power i.e. It should spread evenly on surface and remove dirt.

7] It should have good emulsificability to remove fat particles from the surface of milking equipments.

8] It should be good deflocculating agent to remove milk particles from the surface of milking equipments.

9] It should act as good solvent for removal of protein particles from the surface of milking equipments.

10] It should be easily removable from surface of utensil by water use.

11] Its cost should be affordable and available easily.

Types of dairy detergents

a] **Acid cleaners-** e.g. Phosphoric acid, Tartaric acid, Citric acid, Gluconic acid, 1 to 2% solution, pH 9.8 to 12.2.

b] **Alkali cleaners-** e.g. Sodium sesquicarbonate, Sodium bicarbonate, Sodium carbonate, Sodium metasillicate, Trisodium phosphate, Caustic soda, 1% solution, pH 6.5 to 6.8.

c] **Complex phosphates-** e.g. Sodium tetrametaphosphate, Sodium hexa metaphosphate, Sodium tri-poly-phosphate. 1 to 2% Solution, pH 7.5.

d] **Wetting agents-** e.g. Teepol

An example of dairy detergent mixture-

Trisodium phosphate	40 parts
Sodium carbonate	40 parts
Sodium silicate	20 parts

Sterilization

Sterilization of dairy utensils can be defined as the total destruction of all pathogenic and non-pathogenic bacteria from the product contact surfaces of dairy utensils and equipments

- It ensures complete destruction of microorganism and their spores.
- It is done by use of following
  1] Chloride gas
  2] Calcium hypochlorite powder [Aqueous solution with Chlorine availability of 200 PPM]
  3] Sodium hypochlorite [Aqueous solution with Chlorine availability of 200 PPM]
  4] Chloramine-T
  5] Steam at 15 Ibs. pressure

Precautions

- Acid solution should not be used for cleaning aluminium utensils.
- After the use of acid cleaner alkali solution should be used to neutralize acidic effect on metal/milk.
- During cleaning of utensil hot water must not be used at the first step as it would form clots of protein on the surface which leads to formation of milk stones.
- Before sterilization of utensils, their proper cleaning is very important.

Steps in cleaning and sterilization of dairy utensils and equipments

The important steps in cleaning and sterilization of utensils and equipments are described as below,

1. **Draining of utensils and equipments:** This is the first step in cleaning of utensils and equipments. In this step the residual milk or liquid present in utensils and equipments is completely drained out by inverting the utensils or equipments.
2. Free rinse of utensils and equipments with cold water or tap water or cold water rinse.

   In this step after the complete drainage of liquid milk from the utensils and equipments. The thin film of milk on surfaces of utensils and equipments can be removed by using tap water.
3. **Hot detergent cleaning:** In this step detergent solution of 1-2% washing soda with 12.2 pH is used effectively for cleaning of insides and outsides of utensils and equipments. While cleaning scrubbing brush should be used by the operator to accelerate the cleaning action here the temperature of detergent should be kept 55 ^{0}C. In this step the thin film of milk present on surfaces can be removed completely.
4. **Hot water cleaning:** In this step again the utensils and equipments are cleaned with the help of hot water so as to remove traces of detergent solution present on the surfaces of dairy utensils and equipments.
5. **Sterilization of utensils and equipments:** After cleaning the utensils and equipments just before their use the surfaces of utensils and equipments can be sterilized by using either boiling water or steam for this purpose chemical sanitizer like chlorine solution of 200 ppm strength can be also used effectively.

This process helps to destroy all the pathogenic and non-pathogenic bacteria from the surfaces of the utensils and equipments.

6. **Draining and drying of utensils and equipments:** After the cleaning and sterilization of utensils and equipments all the utensils are placed on the draining rack in inverted position. This practice will help to prevent the bacterial growth and corrosion of surfaces. This process can be accelerated by the proper use of ventilation. Never use cloths for drying of utensils and equipments because such practice will cause the contamination of surfaces and later on it may cause the spoilage of milk and milk product.

   Note:

   - If chlorine solution as a sanitizer is left in contact with surfaces it will cause the corrosion of metallic surface. Hence, its contact period should be keep minimum.
   - Periodically acid cleaner solutions such as 1% phosphoric acid, tartaric acid with 6.5 pH should be used for cleaning the milk utensils. If cleaning of utensils is not done properly then milk solids are left on the joints or the surface of utensils it becomes the source of bacterial contamination. Continuous deposition of milk solids will form the milk stones which later on spoils the surfaces of utensils.
   - Washing the utensils with cold water in step first to flush of as much milk solids as possible. Use of hot water initially will helps to form the clots of milk protein on the surfaces which later on form the milk stones.

7. Use of alkali solution (washing soda) in the cleaning helps to neutralize the lactic acid present in the milk and it will accelerate the removal of milk solids while brushing the surface of utensils.

Milk Stones

- When utensils are not cleaned properly or hot water is used in initial step for rinsing milk solids/clots are not removed properly from the surface.
- Accumulation of salts of washing solution and salts of hard water with milk solids form milk stone which is very ideal site for multiplication of bacteria.
- Continuous deposition of milk stones spoils the surface of utensil and contaminates the milk contained in it with harmful microbes.
- In case of formation of milk stones, 0.5 to 1% Sulphuric acid or weak acid like acetic acid used for cleaning the surface.

CHAPTER 29

EMBRYO TRANSFER AND THEIR ROLE IN ANIMAL IMPROVEMENT

What is embryo transfer?

It is the collection of embryo from a superovulated donor cow and transferring the embryos to synchronized recipients to complete the geststion period is called embryo transfer.

Selection of donor cow

1. Top quality female free from reproductive abnormalities, genetical defects, proven maternal qualities should be select as a donor cow.
2. Personal goals/preferences play large part in selection of donor.
3. Marketability of donor calves.

Superovulation of donor cow

- Use Follicle stimulating hormone for superovulation in the donor cow.

Superovulation of donor cow on day 0 to day 4

- Inj. 2 ml on day 0 and day 4 (day 0 starts from 8-14 days following estrus cycle).
- Give Inj. Prostaglandins on day 3 to donor cow for superovulation.
- On day 5- onset of estrus in donor cow.

Inseminate the superovulated donor cow from 5th day when there is standing heat.Usually multiple inseminations on 12, 24, and 36 hours after onset of estrus should be carried out in donor cow. Preferably more than one straw of more quality semen should be used to inseminate the donor cow.

Embryo recovery from donor cow

Recovery of embryo should carried out approximately seven days of breeding. For recovery of embryo only 30 minutes time is required. Insert foley catheter with inflatable ballon into donors uterus. Then introduce flushing solution in each horn of uterus. Gently massage the solution filled uterine horn with hand. Fluid containing the embryo is drawn backout and collected through a filter into a holding cyllinder.

Embryo processing

After 30 minutes of embryo collection embryos located with stereoscopic microscope. Embryos are washed and transferred to holding medium. This procedure generally will be carried out for three times.Then embryos are eveluated for state of development and quality. Firstly classified the embryos as good and bad. Then good embryos are further classified.

Selection of Recipient

1. Recipient cow have sound and good mothering ability.
2. Recipient should have good health and nutrition.
3. Must be synchronized to receive embryos.

METHODS OF SYNCHRONIZING OESTRUS

- The most widely accepted procedure for synchronizing recipients is administration of a luteolytic dose of prostaglandin F_2alpha or a suitable analogue during the luteal phase. This is probably superior to using natural oestrous cycles.
- Injecting potential recipients with two doses of prostaglandin at 11-day intervals when stages of the reproductive cycle are unknown also works well if cattle are cycling.

Embryo transfer

Two methods of embryo transfer

1. Surgical method &
2. Non-surgical method

1. Surgical transfer

- mid-line abdominal incision to cows under general anaesthesia
- Left flank incision is far more practical.
- The CL is located by rectal palpation.
- the flank ipsilateral to the CL is clipped, washed with soap and water, and sterilized with iodine and alcohol.
- About 60 ml of 2 percent procaine is given along the line of the planned incision.
- the surgeon makes a skin incision about 15 cm long, high on the flank, just anterior to the hip
- Muscle layers are separated, and the peritoneum is cut.
- The surgeon inserts a hand and forearm into the incision, locates the ovary, usually about 25 cm posterior to the incision, and visualizes or palpates the CL.
- The uterine horn is exteriorized by grasping and stretching with the thumb and forefinger the broad ligament of the uterus, which is located medial to the uterine horn.
- A puncture wound is made with a blunted needle through the wall of the cranial one-third of the exposed uterine horn.
- Using about 0.1 ml of medium in a small glass pipette (<1.5 mm outside diameter), an assistant draws up the embryo from the storage container.
- The pipette is then inserted into the lumen of the uterus, and the embryo is expelled.
- The incision is then closed, using two layers of sutures. With practice, the surgery takes about 15 minutes.

2. Non-surgical transfer

- First, it is necessary to be able to palpate ovaries accurately in order to select the side of ovulation.
- Pregnancy rates are markedly lowered if embryos are transferred to the uterine horn contralateral to the corpus luteum.
- Also, recipients should be rejected if no corpus luteum is present or pathology of the reproductive tract is noted.
- The next step is to pass the embryo transfer device through the cervix.

- During luteal phase embryo transfer is difficult than the estrus phase because of limiting opeing of cervix
- The third step with non-surgical transfer is to be able to insert the tip of the instrument into the desired uterine horn quickly, smoothly and atraumatically.
- Well-trained inseminators generally require 100–200 non-surgical transfers until their pregnancy rates plateau; others usually require more.
- Most technicians who are successful with non-surgical transfer had low pregnancy rates for their first 100 non-surgical transfers.
- If one begins with non-surgical transfer and pregnancy rates are low, it is difficult to distinguish among problems such as identifying usable embryos, problems with media, problems in storing embryos from collection to transfer, poor non-surgical embryo transfer technique, recipient problems, etc
- Some people believe that there is a 5–10 percent advantage in pregnancy rates with surgical transfer, even when very proficient technicians are doing the non-surgical transfer.

  Even if this is true, in most circumstances nonsurgical transfer is still preferred because it is less expensive; it is quicker and does not involve surgical procedures.

Loading of straw

- The first step is to take a sterile 0.25-cc straw, shortened by 1 cm before sterilization, label it and rinse it twice with medium to removed any toxic contaminants, taking care not to wet the cotton plug and to discard the rinses. A plastic 1-cc tuberculin syringe fits snugly over the straw for aspirating and expelling fluid.
- The straw is filled nearly one-third full of fluid, then with a 5-mm column of air, then another column of fluid containing the embryo, one-third the length of the straw, then another short column of air, and finally more fluid to fill the straw and wet the cotton plug. Care must be taken not to compromise the sterility of the tip of the straw or the internal surfaces.

ANAESTHESIA

- Epidural anaesthesia is recommended for routine non-surgical transfer
- This relaxes rectal musculature, making it easier to manipulate the reproductive tract gently as is required for high pregnancy rates.
- Give the epidural injection about five minutes before embryo transfer while the technician is transferring the embryo to the previous recipient.

TRANSFER PROCEDURE

- The actual embryo transfer process is similar to the method used for artificial insemination, except that the transfer gun is passed well up the uterine horn ipsilateral to the corpus luteum.
- A good site to aim at is the palpable bifurcation of the uterine horns.
- The key is to pass the gun without damaging the endometrium. Therefore, it is better to insert the instrument less deeply and not cause damage.
- As with artificial insemination, the plunger should be depressed firmly, but not too rapidly.
- Procedure will be fast without damaging the endometrium.

Advantages of Embryo transfer (ET)

1. Increased number of calves of genetically superior cows.
2. Increased marketing opportunities-offspring & embryos.
3. Ease of import/export.
4. Embryos can be stored indefinitely.
5. Twinning may be possible.

Disadvantages of ET

1. Increased expenses & higher break-even costs for calves.
2. Estrus detection required.
3. Synchronization of recipient with donor is required.
4. Specialized equipments & trained personnel required.
5. More expensive & time consuming than traditional reproductive methods.

CHAPTER 30

INTRODUCTION TO BIOTECHNIQUES IN DAIRY ANIMAL PRODUCTION

Biotechnology

The use of living organisms or their products to enhance the lives & environment.

Biotechniques used to enhance the production

1. Artificial insemination
2. Embryo transfer
3. Embryo sexing & cloning
4. Hormone use
5. Multiple ovulation embryo transfer and open nucleus breeding system
6. Genetic characterisation of animal genetic resources
7. Disease diagnosis
8. Vaccines
9. Increasing digestibility of low-quality forages
10. Improving nutritive value of cereals
11. Removing anti-nutritive factors from feeds
12. Improving nutritive value of conserved feed
13. Improving rumen function

3. Embryo sexing & cloning

- It has been suggested that, if multiple sexed-embryo transfer became as routine an operation as AI is, beef operations based on this system could become competitive with pig and poultry production in terms of efficiency of food utilisation.

- Clones may be produced by embryo splitting and nuclear transfer.
- cloning of embryos will not increase rates of genetic progress in the nucleus, but that it offers considerable advantages in increasing the rate of dissemination of tested superior genotypes in commercial populations.
- Other potential applications of cloning include efficient evaluation of genotype x environment interactions and testing and/or dissemination of transgenics.

4. Hormone use

- Use of hormonal assays to *monitor* reproductive function can be rewarding for both research purposes and commercial livestock operations. Reproduction can also be *manipulated* using hormonal treatments.
- Hormones produce desirable effects in animals.
- Lack of Knowledge about use of hormones & less economic viability of hormones limits the utility.
- Use of hormones leads to increase ovulation rate, induction of estrus, improves fertilization rate etc.

5. Multiple ovulation embryo transfer (MOET) and open nucleus breeding system

- Multiple ovulation embryo transfer (MOET) is a composite technology which includes superovulation, fertilisation, embryo recovery, short-term *in vitro* culture of embryos, embryo freezing and embryo transfer.
- Benefits of MOET
  1. Increasing the number of offspring produced by valuable females.
  2. Increasing the population base of rare or endangered breeds or species.
  3. *ex situ* preservation of endangered populations.
  4. Progeny testing of females.
  5. Increasing rates of genetic improvement in breeding programmes.

  Open nucleus breeding system
- The ONBS concept is based on a scheme with a nucleus herd/flock established under controlled conditions to facilitate selection.
- The nucleus is established from the "best" animals obtained by screening the base (farmers) population for outstanding females.

- These are then recorded individually and the best individuals chosen to form the elite herd/flock of the nucleus. If ET is possible, the elite female herd is used through MOET with superior sires to produce embryos which are carried by recipient females from the base population.
- The resulting offspring are reared and recorded and the males among them are evaluated using, as appropriate, the performance of their sibs and paternal half sibs and their own performance.
- From these, an elite group of males with high breeding values for the specific trait is selected and used in the base population for genetic improvement through natural service or AI.

6. Genetic characterisation of animal genetic resources

- Some livestock breeds in these countries are in immediate danger of loss through indiscriminate crossbreeding with exotic breeds.
- However, most of these animal genetic resources are still not characterised and boundaries between distinct populations are unclear.
- Randomly amplified polymorphic DNA (RAPD), polymorphism at the level of DNA are used for genetic characterization of animals.

7. Disease diagnosis

- Successful control of a disease requires accurate diagnosis.
- Monoclonal antibodies are used to immortalise individual antibody-producing cells by hybridisation to produce antibodies of a given class, specificity and affinity.This technique has provided a tool that permits the analysis of virtually any antigenic molecule.
- ELISA (enzyme-linked immunosorbent assays)
- The ability to generate highly specific antigens by recombinant DNA techniques has made it possible for an increasing number of enzyme-linked immunosorbent assays (ELISA) to have the capacity to differentiate between immune responses generated by vaccination from those due to infection.
- The advent of PCR has enhanced the sensitivity of DNA detection tests considerably.
- For example, PCR used in combination with hybridisation analysis, has been shown to provide a sensitive diagnostic assay to detect bovine leukosis virus.

8. Vaccines

- Immunisation remains one of the most economical means of preventing specific diseases.
- An effective vaccine can produce long-lasting immunity. In some cases, vaccination can provide lifetime immunity. Moreover a small number of doses is usually required for protection.
- Most of the vaccines are produced by recombinant DNA technology.
- Level of infrastructure and logistical support required for a large-scale vaccination programme is such that a successful vaccination campaign can be implemented in remote rural areas.
- In general, vaccines offer a substantial benefit for comparatively low cost, a primary consideration for developing countries.

9. Increasing digestibility of low-quality forages

- Low-quality forages are a major component of ruminant diets in the tropics. Thus, much progress can be made by improving the forage component of the ration. The characteristic feature of tropical forages is their slow rate of microbial breakdown in the rumen with the result that much of the nutrients of the feed are voided in the faeces.
- The slow rate of breakdown also results in reduced outflow rate of feed residues from the rumen which consequently depresses feed intake. At present, the main treatment methods for forages such as cereal straws are either mechanical (e.g. grinding), physical (e.g. temperature and pressure treatment) or a range of chemical treatments of which sodium hydroxide or ammonia are among the more successful.
- The lignification of the cell walls prevents degradation by cellulase or hemicellulase enzymes.
- Fortunately, it is possible to use lignase enzyme produced by the soft-rot fungus.

10. Improving nutritive value of cereals

- Moderate protein content and low amounts of specific amino acids limit the nutritive value of cereals and cereal by-products (e.g. barley is low in Iysine and threonine).
- This is a major limitation in the ration formulation for non-ruminant livestock which necessitates addition of expensive protein supplements.

- There are on-going studies to enhance the low level of Iysine in barley by genetically engineering the grain genome.

11. Removing anti-nutritive factors from feeds

- Anti-nutritive factors in plant tissues include protease inhibitors, tannins, phytohaemagglutinins and cyanogens in legumes, and glucosinolates, tannins and sanapine in oilseed rape (*Brassica napus*) and other compounds in feeds belonging to the *Brassica* group.
- Transgenic rumen microbes could also play a role in the detoxification of plant poisons or inactivation of antinutritional factors.

12. Improving nutritive value of conserved feed

- The conservation of plant material as silage depends upon anaerobic fermentation of sugars in the material which in turn is influenced by the ability of naturally occurring lactic acid bacteria to grow rapidly on the available nutrients under the existing physical environment.
- Throughout this century, research workers have investigated ways through which the fermentation process in silage making can be controlled in order to improve the feeding quality of the resulting silage.
- Enzymes are essential for the breakdown of cell-wall carbohydrates to release the sugars necessary for the growth of the lactic acid bacteria. Although resident plant-enzymes and acid hydrolysis produce simple sugars from these carbohydrates, addition of enzymes derived from certain bacteria, e.g. *Aspergillus niger* or *Trichoderma viridi* (Henderson and McDonald, 1977; Henderson et al 1982) increases the amount of available sugars.
- Commercial hemicellulase and cellulase enzyme cocktails are now available and improve the fermentation process considerably (Hooper et al 1989). However, prices of these products preclude their viability for farm level application, especially in developing countries.
- Commercial bacterial inoculants designed to add sufficient homofermentative lactic acid bacteria to dominate the fermentation are now available The objective of using such additives is to ensure the rapid production of the required amount of lactic acid from the carbohydrates present to preserve the ensiled material.
- Most such inoculants contain *Lactobacillus plantarum* with or without other bacteria such as *L. acidophilus, Pediococcus acidilactive* and *Streptococcus thermophilus*

13. Improving rumen function

- Improvement of rumen function by transgenic technology.
- These include development of transgenic bacteria with enhanced cellulotic activity, capability to cleave lignohemicellulose complexes, reduced methane production capability decreased proteolytic and/or deaminase activities, increased capability for nitrogen "fixation" and increased ability for microbial production of specific amino acids.

CHAPTER 31

DEMONSTRATION OF SEMEN COLLECTION, PROCESSING

Objectives

1] To be acquainted with the semen collection procedure, processing of the collected semen and artificial insemination [AI].

2] To be acquainted with the advantages of AI over natural service.

1] Semen collection

Following are the methods of semen collection in case of AI in bovines-

- The massage method- Arm is inserted into the rectum of bull and ampullae are massaged, Particularly, this method is used in the bulls with injury to legs and is unable to mount, Less quantity of semen is obtained in this method.
- Ejaculation by electrical stimuli- Used in sluggish bull and the bull which is unable to mount.
- Collection of semen from vagina after natural service- This collection is done by spoon or rubber tube or aspirator. Being mixed with secretions from female genital tract, the semen collected this way has poor keeping quality.
- Artificial vagina method [AV method]- It is most widely used method of semen collection. AV set is made up of following items/parts-

  i] Heavy rubber cylinder [Two feet long and six and half cm. Diameter]

  ii] Rubber sleeve or rubber liner.

  iii] Semen receiving cone and rubber

  iv] Graduated vial for collection of semen

Method of collection of semen using AV method

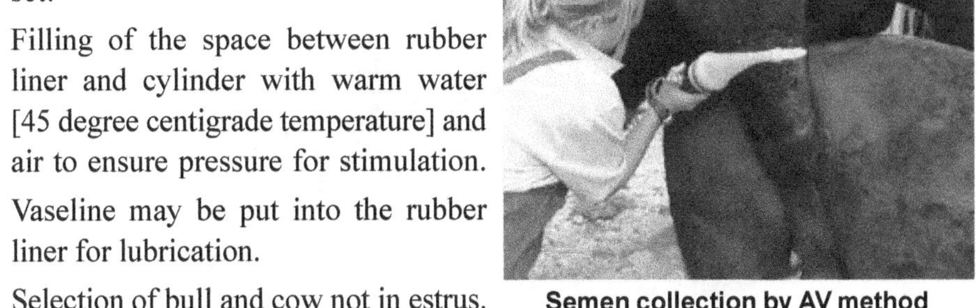

Semen collection by AV method

1] Sterilization of all the parts of the AV set.
2] Filling of the space between rubber liner and cylinder with warm water [45 degree centigrade temperature] and air to ensure pressure for stimulation.
3] Vaseline may be put into the rubber liner for lubrication.
4] Selection of bull and cow not in estrus.
5] Take cow in crate.
6] Mounting by bull without allow for smelling genitalia of cow.
7] While mounting, sheath is supported and penis is quickly directed in AV set.
8] Semen in collecting vial [Hold AV set in vertical upright position after the mounting ends]

2] Processing

This included-

A] Examination of collected semen
B] Dilution
C] Freezing

A] Examination- For assessment of the quality of semen collected the sample subjected to following tests-

1] Mass activity
2] Motility
3] Colour
4] Consistency
5] Live sperm count
6] Abnormal sperm count
7] Volume
8] Methylene blue reduction test and
9] pH

B] Dilution

1] Diluter is used as semen extender. Some diluters are as follow-

Egg yolk Phosphate, Egg yolk citrate, Tris extender, coconut milk based diluter, milk based diluter, etc.

2] Addition of antibiotic in diluter for increasing conception rate.

C] Freezing

- Frozen semen preserved at -196 deg. Centigrade in liquid nitrogen in stainless steel container. Semen is packed in ampoules or straws. For thawing of frozen semen water with 35 deg. Centigrade temperature.

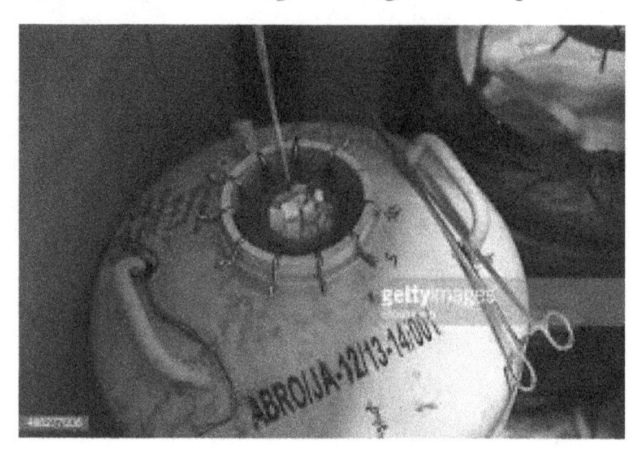

Liquid nitrogen container

CHAPTER 33

HANDLING AND RESTRAINING OF DAIRY ANIMALS

Handling and controlling [restraint] of animals is required during several farm operations like treatment, castration, surgical procedures, breeding, milking, shearing, dipping, examination of mouth, eyes, teeth, hooves, feet, assistance during calving, dystocia, drenching, etc. Handling must be gentle. The person who handles animal should be cautious with regard to his own safety failing to which sometimes it may result in injury or fatality.

Certain instruments, equipments are generally used for handling and restraining of animals some of them are detailed below.

1] **Mouth Gag-** For keeping two jaws apart during examination of oral cavity, for operation, for passing probing or drenching tube. Mouth gags are mainly of two types, a) Drinking water gag & b) Probang gag

a) **Drinking water gag:** These are made up of aluminium. These are two in number. One for upper jaw and another for lower jaw. These correspond to the shape of jaw. Both are fitted with flanges to accommodate cheek teeth.

b) **Probang gag:** It is made up of wood. It has hole in the centre to pass probang. It is also provided with a strap which goes over the poll.

2] **Muzzle-** are used to prevent the cow from eating soil, inanimate objects, self-sucking, to prevent bullocks from grazing in cultivated crop during drafting operations in the fields, etc.

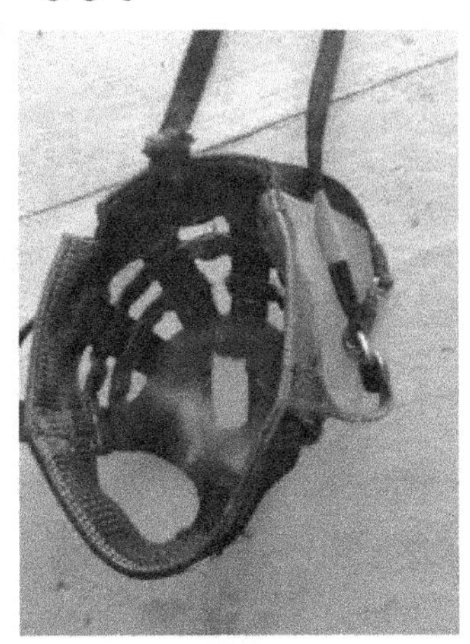

Muzzle

3] **Ring-** Metallic rings used for controlling vicious and aggressive bulls.

4] **Halter-** Device made up of three strand rope, 10 to 12 feet long.The rope is put round the face with one string around the poll/neck.

5] **Bull holder/Bull dogs/Bull tongs/ Bull nose ring-** Metallic device, applied on nostrils when examination needs more time.

Halter and Bull nose ring

The bull nose ring is used for better control of bull. It consists of 2 semicircular pieces hinged together. It is made from non-rusting metal like copper or aluminium. There are two types of nose ring. I) Non-piercing & ii) self piercing type. Each year, nose ring is replaced by bigger size. Bull pole or bull leader is attached to nose ring for control and leading of bull.

6] **Bull mask-** Made up of strong leather or galvanized steel. It rests on the front of face and prevents bull from seeing straight ahead.

7] **Spring gag-** Consists of collar and licking screw device in which sliding arm can be fixed in any position in the oral cavity.

8] **Suckling preventer-** A sort of shield made up of aluminium for preventing suckling and self suckling.

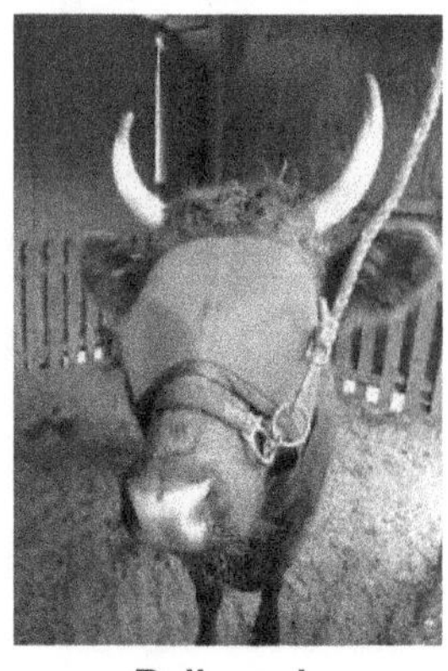

Bull mask

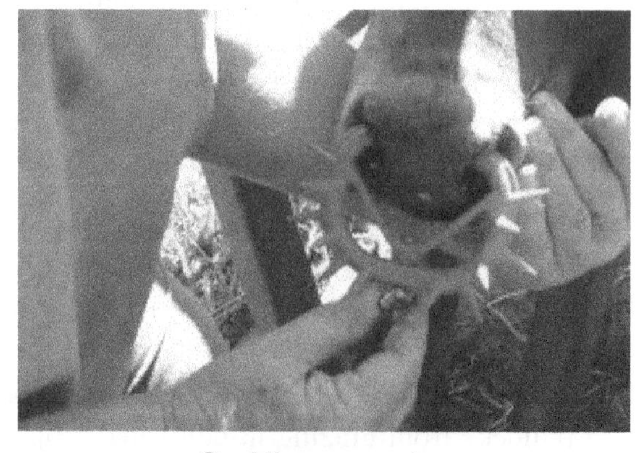

Suckling preventer

9] **Trevis-** These are made up of hard metal tube consist of four poles fix in the ground in addition there are two or more fixed side walls or side rods. The front and hind wall are movable with adujusted size (height) of animal in this way; it is useful in controlling the animals in clinics.

10] **'8' type knot/ anticow kickers-** Used during milking in case the cow has bad habit of kicking. Soft cotton rope is tied to both the hind legs immediately above hock joint before start of milking in such a way that the knot appears as digit '8'.Anticowkicker is made up of two metal clips connected together by chain. The clips are fitted above the hocks with chain hanging infront of hocks. A smaller clip is provided to fix the tail at left clip.

Trevis

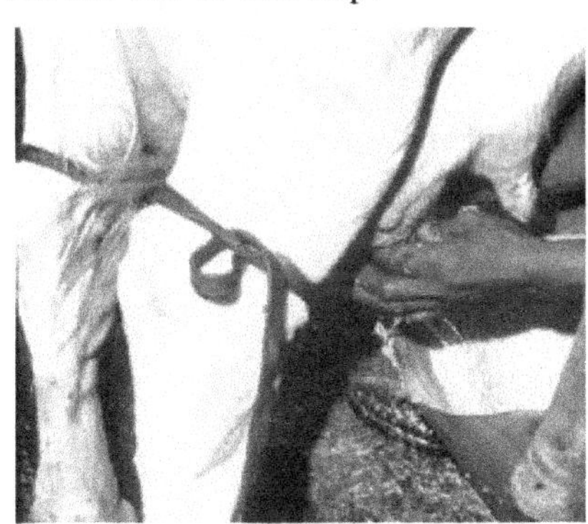

'8' Knot method

Casting /Throwing-

Casting is defined as throwing of animal on the ground slowly and safely.

It may be necessary sometimes to throw and tie animal for certain operations. For this preferably cotton rope of one inch diameter is required the length of which should be 30 to 40 feet.

Methods of casting

1] **Reuff's method of casting**

2] **Abilgaards alternate method of casting**

1] Reuff's method of casting

- Take a long rope [preferably cotton rope].
- Make a noose/knot at one end of rope.
- Pass it round the base of horns [In polled animals place it around the neck].

- Make a half hitch around the neck and a second round at the chest immediately behind the elbow and third round at the abdomen in front of the udder or scrotum.
- Pull the rope with the help of two assistants when the animal will sink to the ground.Pass the tail forward from in between the thighs and pull around the top side of thigh.
- Secure all the feet and tie together or tie the four legs to any fixed object.
- Care should be taken that the animal should not fall on the left side.
- Keep head and neck of the animal down after sinking on ground for minimum struggling and danger of breaking horn.
- Secure all the legs and tie together.
- Use two more ropes for correction of torsion or intestine by rolling the animal one for tying front leg to the hind leg of the same side and other for tying right and left legs together.

Reuff's method of casting

2] Abilgaards alternate method of casting

It is an alternative method of casting cattle. In this method, all the four legs are controlled in the beginning. Large open space is needed for application of this method.

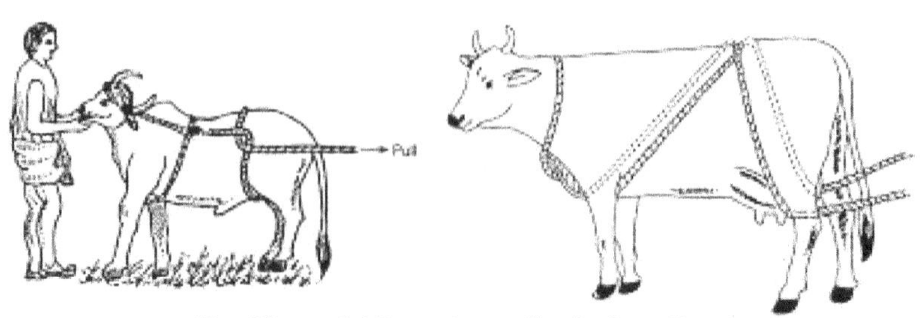

Reuff's and Alternate method of casting

Place the middle portion of rope around the neck. Cross the free ends under neck. Pass the free ends in between the elbow backwards one on either of animal. Pass the free ends over the loin region and cross to the other side. Draw the both ends backwards between the thighs and pass through space between the udder and thigh. Pull the free ends of the rope backwards. Animal then falls slowly on the ground.

Precautions before casting

i] As far as possible, keep the animal off fed at least 12 hours before casting to avoid risk of rupture of abdominal organs due to pressure of feed/ingesta.

ii] The animal should be muzzled during fasting period to prevent eating bedding or other material.

iii] Avoid casting during extremes of weather.

iv] Preferably cast the animal in open space.

v] Animal should always be casted on a smooth, large well straw bedded area or grass covered area.

vi] Advanced pregnant animal should not be cast.

vii] Plenty of assistance should be at hand.

CHAPTER 33

JUDGING OF COWS AND BUFFALOES

Purpose

1] Motivation of farmers for competition.

2] Motivation for development purity of breeds.

3] For selection of animals for breeding, cattle improvement programme.

4] Advertising.

5] Attraction for prospective buyers.

6] For dissemination of knowledge of breed characters.

7] To promote interest in dairying.

The animals in the judging contest are generally categorized breed wise, species wise on the basis of age, sex, physiological status as below-

Calves under one/two year[s] [irrespective of sex]

Heifers/Pregnant heifers

Pregnant cows

Lactating cows

Breeding bulls

Score in cattle judging is based on general appearance, apparent look of animals, their body capacity, development of mammary system, etc. The animal should be true to the breed type. In females, feminine characters should be well developed, in males masculine vigor should be well expressed. All body parts should be harmoniously. Body should be proportionate and should exhibit breed characters.

Udder should be capacious, symmetrical. Pliable, smooth, firm to touch, wide, deep, teats of uniforms size, length, four in number in cows and buffaloes. Well developed, zig zag mammary vein is thought be indicative of dairy characters.

Generally, score card is made up of 100 points and these points are categorized as below-.

Parameter	Maximum score	Subscore points for
General body confirmation	30	Head, jaws, nose, limbs, tail, Buttock, eyes, hooves
Dairy Character	20	Feminine expressions and constitution, udder, teats, mammary vein, neck, face, legs.
Body capacity	20	Chest, ribs, barrel, pin bones, nostrils.
Mammary system	30	Shape of udder, teats milk vain

On the basis of score point received the animals in judging contest are graded as below-

Score range	Grade
90 and above	Excellent
85 to 89	Very Good
80 to 84	Good
70 to 79	Acceptable
60 to 69	Fair
Below 60	Poor

Following body defects be evaluated and graded accordingly-

1] **Slight Discrimination:** Minor lameness, cross eye, teats of unequal size, cropped ears, slight asymmetry of udder, minor wound [s].

2] **Medium Discrimination:** Wry face turned to one side, parrot jaw, blind with one eye, weak/ broken attachment of udder, lack of size.

3] **Serious Discrimination:** Blind with both eyes, free martin heifer, obstruction in teats, fibrosed udder, blind quarter, lameness with hind limbs, wedge shaped udder, hard lump on udder, over conditioned body, sign of operation to conceal, evidence of milking uncalved in case of heifer.

Coat score in cattle

Coat Score	Coat Type
1	Extremely Short
2	Very Short
3	Fairly Short
4	Fairly Long
5	Long
6	Woolly
7	Very Woolly

Score card for judging of cattle & buffaloes

	Subscore	score for animals (breed)						
a) General appearance (Total score 30)								
1. Breed characteristics, head	10							
2. Shoulder blades, back, rump	10							
3. Legs and feet	10							
b) Dairy characters (Total score 20)								
1. Neck, withers, ribs, flanks	10							
2.Thigh, skin	10							
c) Body capacity (Total score 20)								
1.Barrel	10							
2. Heart girth	10							
d) Mammary system (Total score 30)								
1. Shape of udder	10							
2. Fore udder	06							
3. Rear udder	07							
4. Teats	05							
5. Milk veins	02							
Total score for animal								
Rank of the animal								

CHAPTER 34

PREPARATION OF ANIMAL FOR SHOW

Different animal shows/ exhibitions are organized by various institutions at different places of India. The objectives behind the organizing animal shows/ exhibitions includes to create awareness among the farmers regarding different breeds of animal, indigenous breeds of animal, to make market available for buying and selling of animals, to make aware the different physical characteristics of different breeds of animals etc.

Certain steps involved in preparation of animal for show are as follows,

Selection of animal, early preparation of animal and preparation on the day of show.

Selection of animal

Animals having following characters should be selected for atleast 2-3 months before the show.

1. True breed characters
2. Good dairy characters
3. Good temperament
4. Proper growth and development according to age
5. Free from physical defects or deformities

Early preparation of animal for show

1. Animal should be given an additional amount (approximately 1-1.5 kg) of concentrate mixture daily.

2. Animals should keep in comfortable, cool, dry and well ventilated house provided with smooth bedding.
3. A moderate exercise should be given daily to animal.
4. The animal should be trained to lead/ For leading hold the rope halter near to head and keep the head straight while walking and standing.
5. Animal should be trained to stand squarely on its feet with the front feet apart and one back foot slightly behind the other.
6. Hooves should be trimmed well and shaped properly.
7. Hairs on the belly and udder should be clipped for distinct look of milk and udder veins. Also clip hairs from switch of tail to give proper shape.
8. Grooming should be undertaken daily so as to make hair glossy.
9. Animal should be washed every week to remove dirt and loose hairs.
10. Body coat of animal should be rubbed vigorously with piece of cloth or blanket so as to give glossy look to skin.

Preparation on the day of show

1. Wash the animal body with soapy water and dry it with towel.
2. Rub the horns and hooves with sand paper and then apply coconut oil.
3. Comb the tail switch to give clean and fluffy look.
4. Groom the animal with a soft brush and rub with soft cloth.
5. In animals with white body coat use talc powder to increase the luster of body coat.
6. A day before show animal is half fed so that on the day of show animal takes large quantity of food. This helps to show bigger body capacity.
7. Animal is make thirsty a few hours before the timing of show by restricting its water intake and then animal is allowed to drink the water *ad-lib*. This is called filling which also helps to exhibit better body capacity.
8. The animal should not be milked 6-8 hrs. prior to show to exhibit better capacity of udder and well placed teats.
9. Lead the animal in show ring with the help of leather halter and clean rope.
10. Keep the animal calm and comfort.
11. Cover body of animal with clean cloth.

CHAPTER 35

SILAGE MAKING

It is the method of controlled anaerobic fermentation of green fodder followed by storage in specially prepared silo. The process of silage making is called ensiling.

Suitable fodder crops for silage making

The best silage can be prepared from cereal fodder crops like Jowar, Maize etc. The silage can also be prepared from sugarcane leaves, whole sugarcane, natural grasses, perenial grasses, legumes by addition of certain additives in silage.

Different types of silos

1. Tower silo-(6 m height and 3 m diameter or 9 m height and 6m diameter)
2. Trench silo-cement concrete
3. Pit silo- Height of pit/Tower of silo should not exceed more than 9 meters.

Method of silage making

Harvest the crop at preflowering stage in maize, milk or dubt stage in Jowar & Bajara. The harvested fodder is allowed to dry for some time so as to reduce the moisture content to 65%. Then it is chaffed with the help of chaffcutter. At the bottom of silo it is better to spread a layer of hay or straw. The chaffed fodder material be evenly spread throughout the silo and pressed to expel the maximum airout. While filling silo it is better to complete the process continuosly in shortest period of time and not to prolong it for over many days. Raise the level of ensiled mass 1.5 m above the ground which would get reduced in due course of time. Cover the silo with a layer of grass and mud. Keep it for minimum 45 days if only jaggery and salt used & then use for feeding of animals whenever needed. Keep it for 22 days airtight if silage culture is used.

Characteristics of good silage: It has good aroma. pH varies between 3.4 to 4.2. It is greenish or brownish yellow in colour. It has pleasant acidic test.

Advantages of silage

1. Minimum loss of nutrients compared to other method of preservation.
2. Used as a succulent green fodder.
3. Succulent feed is available during lean period & scarcity.
4. Silage can be prepared during all seasons.
5. Succulent feed is made available round the year.
6. Silage requires less area for storing compared to hay.
7. Fields are made available for early preparation of rabi crops because of early harvest of green fodder for silage.
8. It is palatable and laxative feed.
9. Ensiled feed is more digestible than fresh green fodder.

Different methods of preserving silage

1. **Additives-**
   a) Readily fermentable carbohydrates e.g. molasses, whey, butter milk, sour milk etc.
   b) Less readily fermentable carbohydrate supplements e.g. Ground barley, oat meal, Jowar meal, Potato, sugar beat pulp etc.
2. **Chemical preservatives**
   a) Acids- Oxalic acid, formic acid, Lactic acid, Phosphoric acid, Dil.HCL etc.
   b) Chemicals- Sodium sulphite, Sod. Bisulphite, Common salt, Zinc bacitracin etc.
   c) Gases- CO_2, SO_2, Formalin etc.
3. **Miscelleneous:** Microbial cultures- e.g. Lactobacillus plantarum, Lactobacillus acidophillus etc.

Conditions requiring use of additives and preservatives

1. **If a fodder crop:**
   a) is rich in crude protein
   b) Has high moisture content
   c) Has low fermentable carbohydrates

2. To increase feeding value of silage
3. To reduce dry matter and seepage losses.
4. To prevent decay in case of doubtful technical know how.

Suggested quantity of some additives and chemical used for silage making

Preservative per ton of ensiled mass (kg)	Kind of crops			
	Sorghum & Maize	Oat & Grass	Sorghum+ legume	Berseem and Lucerne
Molasses	Nil	15	30	40
Ground grains	Nil	40	50	60
Liquid whey	Nil	120	170	220
75% Phosphoric acid	Nil	4	6	8
Sodium bisulphite	Nil	3	4	5
Sulphur dioxide	Nil	1/2	1	2
Common salt	Nil	10	15	20
12% formic acid	Nil	2	3	4

CHAPTER 36

NUTRITIONAL MANAGEMENT AND MILK COMPOSITION

As the price of milk in India is based on its fat and SNF % by co-operatives and private milk procuring agencies, there is need to focus on increasing fat and SNF level in milk. It is possible to increase fat and SNF in milk by manipulation or changing the routine feeding pattern. Feeding programmes with balanced protein, energy and supplementation of deficient minerals and vitamins increase milk yield and its composition.

Nutritional factors influencing milk composition

1. **Ration forage and fiber content (Roughage: concentrate ratio):** Acetate is the precursor for the fat production and its result with more of roughage digestion. Low forage rations will result in a depression of the acetate to propionate ratio and subsequent decline in milk fat content. More concentrate feeds to more propionic acid production which results into increase in milk production but low milk fat. Roughage: concentrate ratio should be around 60:40 or 70:30 for the normal fat production. For maximum fat production animals should be encouraged for taking more roughage. If the rations contain less than 50% forage or 19% acid detergent fiber (ADF) there is potential for low milk fat tests.
2. **Particle size of forage:** To avoid the wastage of green fodder and dry fodder it is chopped and fed, but unfortunately, without proper knowledge in chopping technology the Indian farmers are loosing their income considerably. The fat percentage of milk is depending upon the feeding of length of fiber. If the length of the fiber is short the production of propionic acid will be more than acetic acid which inturn reduce the milk fat percentage. It is suggested

that the particle size of chopped forages should be 3/8 inch or greater (about 20 to 25 percent of the particles should be 1 inch in length or longer) as ideal length for quality milk production, it varies depending up on the type of green fodders. Finding out the correct length for chopping is skill of the farmer which can be developing with experience.

3. **Concentrate source and methods of concentrate feeding:** Readily fermentable non-fiber carbohydrates like starch and sugar in the concentrate mixture affects the fat and protein level in milk. Feeding high levels of grains can lower milk fat. This reduction of fat is also due to acidic condition in rumen and reduction of fiber digestion.This is related to the replacement of rapidly digested starch source with a source of digestible fiber. The result should be more favourable rumen acidity for fiber digesting bacteria and a more desirable acetate to propionate ratio.

   In India it is general practice to feed the concentrates mixing with full of water. This is incorrect as it reduces production of saliva. Reduction in salivary secretion leads to sub clinical acidosis and low fat syndrome. However some people claim advantage of getting more soluble nutrients especially soluble protein directly in to abomasum but, we should not forget that micro oraganisms in the rumen are deprived to soluble nutrients inspite of feeding.It will affect their growth and production functions so it is advised that animals should be watered either 1 hour before or after feeding the concentrate as this has an important bearing on digestion of feed materials. Feeding forage /straw should be done about 1 to 2 hours before feeding concentrates to buffer rumen which in turn prevents sub-clinical acidosis as well as improve milk fat content when you feed concentrates rumen pH will be lowered drastically resulting in less acetate production which in turn will affect milk fat percentage on next milking.

4. **Ration energy and protein content:** The nutritional variable most correlated with milk protein content is energy. There usually is a rise in milk protein content when energy level of ration is boosted but these rations may lower milk fat. Decrease in crude protein levels in ration reduces milk protein level and increased true protein content in diet not always increases milk protein. Thus it is important to maintain rumen degradable protein (RDP) and rumen undegradable protein (UDP) in diet. Bypass protein in natural protein supplements and rumen protected lysine and methionine increases milk protein.

5. **Added fats in feed or ration:** The most consistent effect of added fats has been a slight lowering of milk protein content. This may be true only when unsaturated fat sources are used. Use of protected fats or whole oil seeds often will raise milk fat content. Keep total ration fat levels between 5-7%

as desirable level. High levels of unsaturated fats may alter rumen microbeal activity and dramatically depress milk fat content. When feeding fat calcium needs to be increased to 1% of diet-dry matter. Free fats in the rumen can tie up calcium rendering it unavailable for the fiber digesting, acetate producing bacteria.

6. **Underfeeding:** Under nourishment of the cow reduce the casein content and milk yields by underfeeding of TDN, milk protein will be depressed more. Underfeeding causes a slight depression in lactose content of milk. Under feeding leads to body weight loss or thinning of animals. Cows that are thin at calving or loose excessive amounts of weight right after calving normally have depressed milk fat tests.

7. **Feed supplements/additives:** Minerals are acting as co-factors for important enzymes, increase the function of different enzyme system and improve the animal productivity and milk composition. Supplementation of trace minerals in highly bio-available form (organic minerals) increases metabolism of the nutrients absorbed from digestive system. Yeast supplementation increases fiber digestion an increases lactic acid utilizing bacteria in rumen thus resulting to more milk and milk fat production.

CHAPTER 37

IMPROVED VARIETIES OF FODDERS FOR ANIMALS

In any animal related farming the major expenditure is on the feeding of animals. As many of Indian farmers not cultivating improved varieties of fodders, there is increase in expenditure on concentrate feeding to fulfill the nutritive requirement of animals. Increase in concentrate feeding to animals increases production cost of final product from the animal husbandry and which further reduces the profit margin of farmers. Therefore nutritious fodder production is the only way in the hand of farmers to reduce cost of production. Selection of fodder varieties must be done on the basis of their nutritive value, palatability and fodder yield.

Some of the fodder varieties and their cultivation practices are given as follows,

Sr. No.	Particulars	Hybrid Napier grass	Dashrath grass/ hedge lucerne	Sudan grass
1	**Land**	Poor, Medium to Excellent, Good water holding capacity	Poor, medium, excellent, slight salty	Loam or clay soil
2	**Land preparation**	Ploughing and 2 times harrowing. Collection of weeds, waste	Ploughing and 2 times harrowing	2 times Ploughing and 2 times harrowing
3	**Sowing time**	March-September	June-July	Kharip-June-July Summer = Feb - March

[Table Contd.

Contd. Table]

Sr. No.	Particulars	Hybrid Napier grass	Dashrath grass/ hedge lucerne	Sudan grass
4	**Sowing**	75cm x 50cm or 90 x 90cm, 120 x 60 cm.	30-50cm distance	On 35-40 cm distance behind plough
5	**Varieties**	NB21, RBN9, IGFRI-7, CO4, DHN6, DHN10 Pekchong (Supernapier)	—	Sweet sudan (SSG-59-3), Methi sudan, JS-263, HFS-478 and G-69
6	**Seed rate/ Ha.**	20,000 Stumps (on 75 cm x 50 cm distance)	2-3 kg	50-60 kg
7	**Fertilizers/Ha**	Nitrogen-90kg Phosporus-15kg Potassium-30 kg Apply these fertilizers after 5-7 days of sowing	Nitrogen-20kg Phosporus-50kg Potassium-20 kg at the time of sowing and after every 3-4 cuttings apply Nitrogen-20kg Phosporus -50kg	Nitrogen-65kg Phosporus-125kg Potassium-35 kg at the time of sowing
8	**Intercultivation operations**	1-2 time weeding	Weeding after 3-4 weeks of sowing	—
9	**Water management**	Every 10-12 days interval	—	Kharip- if rain is not proper Summer-10-15 days interval
10	**Harvesting**	First cutting after 60-75 days of sowing. Next cuttings every 60 days interval	First- after 70 days, and next subsequent cuttings after every 45-50 days interval	First- after 60-65 days, and next subsequent cuttings after every 40-50 days interval (Total 3-4 cuttings)
11	**Fodder production (Qui./Ha)**	2000-2500 (in 8-9 cuttings)	150-250 (4 cuttings)	300-350
12	**Crude protein content (%)**	NB-21, RBN-9=9-10% DHN6, DHN10=11 to 13% Pekchong (Supernapier)=14-18%	15.00	17.00

Sr. No.	Particulars	Dinanath grass	Stylo	Anjan ghas
1	**Land**	Fertile Loam soil, sandy soil	Wide range of soils	well drained, well in light textured soils (red coloured), fertile loam soil
2	**Land preparation**	Ploughing and 2-3 times harrowing	Ploughing and 2-3 times harrowing	Ploughing and 2 times harrowing
3	**Sowing time**	Just before rainy season	15 June to 15 August	Throughout year
4	**Sowing**	Seed may broadcast or drilled at 1 cm in rows 40-50 cm apart	On 30 cm distance	Root slops or seeds: 50 cm spacing in between rows and 30 cm between plants.
5	**Varieties**	—	Humilis, Hemata Guianensis and Scaba	Mallapo, Buffel, IGFRI NO. 3108, 3132, CAZRI-358 and Marwar Anjan (75)
6	**Seed rate/ Ha.**	1-2 kg	10-12 kg	4-5 kg or 33000 seedlings or rooted slips
7	**Fertilizers/Ha**	Nitrogen-40kg P_2O_5-20kg mixed in soil during land preparation.	Nitrogen-15kg Phosporus-60kg Potassium-20kg	Basal dose: 5 T of FYM alongwith 40 kg N and 20 kg P_2O_5. Top dressing: 20 kg Nitrogen after one month of sowing. In subsequent years the crop is top dressed with 40 kg N+20 kg P_2O_5 as a single dose at the onset of monsoon.
8	**Intercultivation operations**	One weeding	One weeding	One – two weeding

[Table Contd.

Contd. Table]

Sr. No.	Particulars	Dinanath grass	Stylo	Anjan ghas
9	**Water management**	2nd and 3rd Watering on 4-5 days interval and after 10-12 days interval	2nd and 3rd Watering on 4-5 days interval and after 10-12 days interval	12-15 days interval
10	**Harvesting**	Several cuttings during year	First cutting on 65-70 days of sowing and then after 40-45 days interval (5-6 cuttings)	In the first year only one cut is to be taken. From the 2nd year the crop gives 3-4 cuts. First cutting on 60 days of sowing and thenafter 30-45 days interval.
11	**Fodder production (Qui./Ha)**	4500	600-800 (4-5 cuttings)	90-100 (3-4 cuttings)
12	**Crude protein content (%)**	19.90	15.45	11.00

Sr. No.	Particulars	Lucerne	Berseem	Cowpea
1	**Land**	Medium to Excellent	Medium to Excellent	Medium to Excellent with 5-6.5 pH
2	**Land preparation**	Ploughing and 2-3 times harrowing	Ploughing and 2-3 times harrowing	Ploughing and 2 times harrowing
3	**Sowing time**	October-December	October-November	January - August
4	**Sowing**	On 30 cm distance	On 30 cm distance	On 30-40 cm distance
5	**Varieties**	Anand-2, Sirsa-9, Local	Ghardan, Meskawi, Zanshi	EC-4216, UPC-5286, UPC-5287, NP-3
6	**Seed rate/ Ha.**	25 kg	30 kg	50-55 kg
7	**Fertilizers/Ha**	Nitrogen-100kg Phosporus-50kg Potassium-40 kg	Nitrogen-100kg Phosporus-50kg Potassium-40 kg	Nitrogen-25 kg Phosporus-50-60 kg

[Table Contd.

Contd. Table]

Sr. No.	Particulars	Lucerne	Berseem	Cowpea
8	**Intercultivation operations**	One weeding	One weeding	One – two weeding
9	**Water management**	2nd and 3rd Watering on 4-5 days interval and after 10-12 days interval	2nd and 3rd Watering on 4-5 days interval and after 10-12 days interval	12-15 days interval
10	**Harvesting**	First cutting on 65-70 days of sowing and thenafter 40-45 days interval (2-3 years)	First cutting on 65-70 days of sowing and then after 40-45 days interval (5-6 cuttings)	50% flowering stage. 55-60 days of sowing.
11	**Fodder production (Qui./Ha)**	2000-2500 (8-9 cuttings)	600-800 (4-5 cuttings)	300
12	**Crude protein content (%)**	19.90	15.45	15.00

Sr. No.	Particulars	Nutrifeed	Sugargraze
1	**Land**	Any type	Any type
2	**Land preparation**	Ploughing and 2 times harrowing	Ploughing and 2 times harrowing
3	**Sowing time**	March to September	March to August
4	**Sowing**	30 x 25 cm	30 x 15 cm
5	**Varieties**	—	—
6	**Seed rate/ Ha.**	5-7kg	12.5-15 kg
7	**Fertilizers/Ha**	Compost manure- 12500 kg Nitrogen-75 kg Phosporus-62.5 kg Potassium-25 kg After every cut apply nitrogen @ 62.5 kg/ha.	Compost manure- 12500 kg Nitrogen-75 kg Phosporus-62.5 kg Potassium-25 kg After first cut apply nitrogen @ 25 kg/ha.
8	**Intercultivation operations**	—	—

[Table Contd.

Contd. Table]

Sr. No.	Particulars	Nutrifeed	Sugargraze
9	**Water management**	Summer season -7 days interval and other seasons-12 days interval	12-15 days interval
10	**Harvesting**	At 1-1.2 m height. Cut 6-8 inches above the ground level to promote faster re-growth for multi-cut. Harvest fodder at every 35-40 days interval & gives 5-8 cuttings.	Two-four cutting at 80 days interval
11	**Fodder production (Qui./Ha)**	625-750 (for every cut)	625 -750 (for one cut)
12	**Crude protein content (%)**	14-16%	11.6%

CHAPTER 38

HYDROPONICS TECHNIQUE FOR FODDER PRODUCTION

Hydroponics' means the technique of growing plants without soil or solid growing medium, but using water or nutrient-rich solution only, for a short duration.

Seeds like maize, barley and wheat are used to grow fodder.

METHOD

For 5 crossbred cows.

For hydroponic system maize grain is very economical.

1) Take 10 kgs of maize grain and put that maize grain in a bucket containing water and keep it for 24 hours.
2) Then transfer that maize grain in cotton cloth and keep it for 24 hours.
3) Transfer that sprouted maize grain in plastic tray. The grains are spread on the specialized growing trays and watered at predetermined intervals with overhead sprays.
4) The grain grows in the same tray for 9-10 days and is ready for harvest at a 10-11 inch height.

The system holds enough trays to ensure you daily desired requirements of feed every day. A feed quality maize grain germinates within 24 hours of seeding.

In one tray 0.5 kgs sprouted maize grain is sufficient. From 0.5 kgs maize grain we get 5.5 kgs green palatable fodder after 9 days. For one crossbred cow (450kgs wt.) 3 trays are used for feeding every day i.e. 16.5 kgs.

Advantages of Hydroponic Technique

1. No soil is needed for hydroponics.
2. The water stays in the system and can be reused thus a lower water requirement.
3. Hydroponic gives green palatable fodder which is extremely high in protein, metabolisable energy which is highly digestible by most animals.
4. Reduces labour requirements (30 mins/ day).
5. Less space requirements i.e. 15ft X 15ft area is sufficient for fodder production for 5 cows.
6. Maize grain is very cheap and gives best results in milk production.
7. The animal will eat the entire mat, roots and green growth. So there is no wastage.
8. Nutritional benefits:-
   i) Greater energy and vitality
   ii) Stimulates the immune response.
   iii) Reduction in antinutritional factors.
   iv) Antioxidant properties.
   v) Rich in vitamins and minerals.
9. No need to wait for rain and soil moisture for plant.
10. No more moving, irrigation lines, ploughing fields, mowing, windrowing, bailing and hammer milling.
11. Low energy consumption (one unit / day).
12. Trouble free operation.
13. No pesticide damage.
14. Ease of harvesting.

Nutritional analysis report of 9 days maize grain fodder

Cost of 1 kgs. 9 days maize grain fodder Rs. 1.80/-

Particulars	Moisture	CP	EE	CF	NFE	Total Carbohydrate	Ca	P
Maize	43.2	14.2	2.9	18	14.8	32.8	0.35	0.37
African Tall	38.2	18.4	1.9	29	2.9	31.9	0.48	0.27

Due to changes in the nutritive characteristics of the fodder (less starch, more sugars, vitamins and lysine) monogastrics such as horses, swine and poultry may have more benefit.

CHAPTER 39

AZOLLA PRODUCTION AND ITS USE IN ANIMAL FEEDING

Aquatic plant, free floating fern Azolla which belongs to the family *Azollaceae* is a good source of protein and it contains almost all essential amino acids, minerals such as iron, calcium, magnesium, potassium, phosphorus, manganese etc, apart from appreciable quantities of vitamin A precursor beta-carotene and vitamin B_{12}. It is also found to contain probiotics and biopolymers. Thus, azolla appears to be a potential source of nutrients and has a considerably high feeding value. The water fern azolla, grows in association with the blue-green algae, Anabaena *azollae*, is considered to be the most promising because of the ease of cultivation, high productivity and good nutritive value. Azolla *pinnata* was used as feed in broiler chicken, laying hens, Juvenile Black Tiger Shrimp, goats and buffalo calves.

Azolla cultivation in pits

Three pits, each having the dimensions 5 m X 4m (20 m2) were made with 0.3 m depth (1ft). Care was taken to see that the floor of the pits were even. All the roots and other unwanted particles were cleared from the pits and precautions were taken such that all.

Corners of the pit were of the same level in order to maintain a uniform water level. Silpaulin sheets of 24 ft X 18 ft were spread out over the pits such that sheets were longer and broader than the pits, with no holes. Sheets were spread out uniformly and the outer edges of the sheets were fixed so that they don't slip down. A thin layer of 10-15 cm soft soil was spread evenly over the sheet of each pit such that no large stones or any other contaminants existed. Later, water was filled to a three fourth level in each pit and regular care was taken to maintain the water upto the same level. About 15 kg of fresh buffalo dung dissolved in 35 liters of water was added into each pit with thorough mixing such that the mixture was spread evenly throughout the area. About 30 g of super phosphate dissolved in 10 liters of water was added to the soil in a zigzag manner. Once the preparation was completed, each pit was inoculated with 5 kg of fresh and pure culture of azolla and water was sprinkled over it. pH of the bottom organic matter and the top water were tested regularly. Once in every 15 days, application of 15 kg buffalo dung, 30 g super phosphate and 30 g of mineral mixture was done to obtain continuous growth of azolla and to avoid nutrient deficiency. In case of contamination of the pits by pests and diseases, a fresh inoculation was done with pure culture of azolla after clearing the previous biomass of azolla and water from the affected pit.

Chemical composition of Azolla on dry matter basis

Sr. No.	Nutrient	%
01.	Dry matter	6.0
02.	Organic matter	75.73
03.	Crude protein	23.00
04.	Ether extracts	3.7
05.	Crude fibre	14.7
06.	Nitrogen free extract	33.84
07.	Total ash	24.26
08.	Acid insoluble ash	7.94
09.	Calcium	2.58
10.	Phosphorus	0.26

CHAPTER 40

FEEDING CARE OF ANIMALS DURING SCARCITY PERIOD

Day by day the water and fodder scarcity for animals is increasing and become a big challenge for animal owners to rear the animals by feeding crop residues and other alternate feedstuffs. As we know that these crop residues are high in fiber content, low digestible, have low nutritive value, less palatable for animals. Due to less palatability of crop residues, 40-50 % of the crop residues go waste due to low intake by animal. So to improve feeding value of crop residues, it is essential to carryout the fodder treatments by means of mechanical, chemical and physical treatment on fodder. Alongwith the fodder treatments it is needed to take due care of the animals while feeding crop residues or only dry fodder to animals for long time.

Care to be taken during feeding of urea treated fodder to animals

1. The treated fodder should keep open before feeding so as to liberate excess ammonia in it.
2. The treated fodder should be fed the ruminant animals above 6 months of age.
3. With the urea treated fodder, 4-5 kg green fodder also to be supplied.
4. Ample drinking water should be provided to animals while feeding the urea treated fodder.
5. Treated fodder should be fed as per requirement of animals.
6. The level of urea should not be increased above recommended level to avoid urea toxicity.

7. Sulphur supplement is essential with urea feeding. Nitrogen to sulphur ratio should be maintained as 10:1.
8. The treated straw or fodder must be gradually introduced / reduced from ration of animals.
9. Antibiotic feeding will not affect urea utilization.

Chemical composition of different crop residues

	Dry grass	Soybean straw	Jowar straw	Wheat straw	Rice straw	Dried sugarcane tops
Moisture%	7.02	2.78	3.57	3.04	3.08	4.42
Dry matter%	92.98	97.22	96.43	96.96	96.92	95.58
CP%	4.41	5.38	2.25	2.84	4.35	1.90
Ether extract%	1.04	1.80	1.28	0.96	1.28	1.28
Crude fiber%	48.75	47.21	39.51	37.03	42.72	33.97
Total ash%	10.01	9.81	6.62	9.50	16.07	7.83
NFE%	36.25	35.80	50.34	49.68	35.57	55.02

Chemical composition of different crop residues after 2% urea treatment

	Dry grass	Soybean straw	Jowar straw	Wheat straw	Rice straw	Dried sugarcane tops
Moisture%	36.14	36.87	35.16	34.51	35.05	40.54
Dry matter%	63.86	63.13	64.84	65.49	64.94	57.79
CP%	4.90	6.42	4.19	3.65	5.59	3.31
Ether extract%	0.54	1.65	1.32	1.00	1.26	0.82
Crude fiber%	35.60	33.73	36.98	31.31	27.58	28.32
Total ash%	9.58	7.59	6.04	10.37	19.93	11.11
NFE%	50.38	50.30	51.47	53.67	45.63	56.44

Chemical composition of different crop residues after 4% urea treatment

	Dry grass	Soybean straw	Jowar straw	Wheat straw	Rice straw	Dried sugarcane tops
Moisture%	36.75	35.02	35.02	35.46	36.28	36.17
Dry matter%	63.25	64.99	64.98	64.54	63.72	64.00
CP%	5.89	7.78	7.63	6.64	9.38	5.10

[Table Contd.

Contd. Table]

	Dry grass	Soybean straw	Jowar straw	Wheat straw	Rice straw	Dried sugarcane tops
Ether extract%	0.85	1.62	1.37	0.81	1.12	1.37
Crude fiber%	35.61	32.0	35.10	29.40	29.35	26.00
Total ash%	9.75	9.96	5.44	9.75	19.33	9.49
NFE%	47.96	48.62	50.46	53.40	40.83	58.04

Due to increased use of more quantity of sole dry fodder or low quality roughages in animal feeding, it reduces the number of rumen microbs and also reduces the activity of rumen microbes.As number of rumen microbs and their activity is reduced the digestibility and utility of these low quality roughages is hampered. It results into deficiency or metabolic diseases in animals. Therefore during scarcity period to increase digestibility and utilization of dry fodder or low quality roughages it is essential to provide the ready energy source to animals alongwith the low quality feedstuff. It will improve the rumen microbial count and also activity of ruminal microorganisms.Use 250 gm to 0.5 kg jaggery or mollases or 0.5 to 1 kg of ground grains like maize, jowar , Bajra in animal ration alongwith the low quality roughages.

Deficiency of nitrogen and sulphur during scarcity period results into reduced or limited activity of rumen microbs. That results into reduced digestibility and utilization of roughages.Therefore to maintain the activeness of rumen microbs it is essential to provide nitrogen through non-protein nitrogenous substances and sulphur as per need.

Sole feeding of dry fodder to animals for longer duration results into vitamin A deficiency. Vitamin A deficiency results into continous lacrimation, reduced immunity of animals, diarrohea, cataract, low absorption of digested nutrients in body etc. To avoid the Vitamin A deficiency during scarcity period (due to lack of green fodder) use alternate suppliments of vit.A in the ration of animals.

During scarcity period mineral and salt requirement of animal body is not fulfilled through available feed and fodder. Mineral and salt deficiency in animal body will results into reduce water balance in animal body, reduced level of essential body secretions and hormones, reduced or limited activity of rumen microbs etc. Further it will results into deficiency diseases, loss of production and animal health. Therefore it is essential to provide mineral bricks in animal houses, regular use of mineral mixture and salt in animal ration to avoid the consequences.

Steam treatment on paddy straw, maize straw or wheat straw will improves the digestibilty of these crop residues and increases availability of protein in the crop residues to animal body.

Use of probiotics in animal ration results into increased number of essential rumen micro organisms and increased activity of rumen microbs.

Advantage of Probiotic feeding to animals

1. Improves feed intake, feed efficiency, growth rate, nitrogen balance, nutrient digestibility and milk production.
2. Create favourable environment to enhance digestion and stimulate immunity.
3. Reduce serum cholesterol level and viscocity of chyme.
4. Influence the activity of rumen microbs, stabilize the rumen environment and prevent gas formation, intestinal infections.
5. Favour rapid adaptation to solid food by pre-ruminant calf.

Use of enzyme mixture in the ration of animals will improve digestibility of available feed and fodder, improves digestive environment in the rumen, improves utilization of feed and fodders, increases availability of nutrients to animal body during scarcity period.

Bypass protein and bypass fat provides more bioavailable protein and energy to animal body which is required to fulfill the protein and energy demand of animal body while feeding the crop residues to animals during scarcity. Usually the nutritive requirement of the animal can not satisfied with feeding of dry fodder or available crop residues to animal. Hence, use of bypass protein and bypass fat will definitely benefit the animal body in terms of maintaining the animal health and production. In general use 30 gm bypass protein and bypass fat twice a day in the ration of indigenous cattle and buffaloes whereas use 50-75 gm of bypass protein and bypass fat twice a day in the ration of crossbred cattle and high yielding buffaloes.

To increase the utilization of crop residues in the feeding of animals, make powder of the crop residues and mix it with concentrate mixture and then fed to animals by preparing pellets of the mixture. For milking animals pellets containing 60 percent crop residue powder with 40% of concentrate mixture and fed to animal upto 10-15 kg as per requirement of animal. For dry animal pellets with 80percent crop residue powder + 20 percent concentrate mixture is used for feeding. For growing calves and kids pellets containing 40-50 percent crop residue powder + 50-60 percent concentrate mixture. These pellets increases utilization and digestion of crop residues also avoids wastage of feedstuff.

Oxalate content of tropical grasses, Bajra, Gajraj, hybrid Napier, Guinea grass, sugarcane leaves in high as compared to other fodders. During early stages of growth, there is a rapid rise in oxalate content followed by a decline in oxalate levels as the plant matures.The oxalate content in napier grass is directly related to the thickness of the stem: the thicker the stem the higher the oxalate content. Therefore, harvesting of fodder at proper stage is important.Young plants contain more oxalate than older plants. Most of the parts of Maharashtra sugarcane leaves /tops are regularly used for feeding of animals which contains more oxalate. In drought situation the green or dry sugarcane leaves is the only roughage available in large quantity for feeding of animals.

Effects of oxalate

Oxalic acid is one of the number of anti-nutrients found in forage. It can bind with dietary Calcium (Ca) or Magnesium (Mg) to form insoluble Calcium or Magnesium oxalate, which then may lead to low serum calcium or Magnesium levels as well as renal failure because of precipitation of these salts in the kidneys. If large quantities of oxalate rich plants eaten, the rumen is overwhelmed and unable to metabolize the oxalate and oxalate-poisoning results. Oxalate poisoning may leads to increase blood pressure in animals and kideny stone in bullocks and bulls.

Treatment to reduce oxalate content

1. Urea treatment on these fodders reduces oxalate content.
2. Soaking of paddy straw in water reduces oxalate contents.
3. Washing of bajra fodder in water two times for 30 minutes reduces oxalates upto 70 percent of feed.
4. Lime stone water treatment on oxalate rich fodder.

To reduce ill effects of oxalate rich fodder use leguminous fodder or tree leaves or 15 gm of lime or bone meal alongwith oxalate rich fodder.

QUESTIONS

Fill in the Blanks

1. The colostrum should be fed to calf within....... hrs. of birth.
2. .... is the synoname of Sahiwal cattle.
3. A system of breeding between very closely related animals is called as ....
4. Canine/incisor teeths present in .... jaw of cattle
5. Feedstuffs provided to animals for 24 hrs. is called...............
6. Process of parturition in cows & buffaloes is called as ....
7. .... is the first solid feed given to calves.
8. Estrus period in cow lasts from .... to ..... hrs.
9. India ranks .... in total milk production in world.
10. The time interval between date of drying off the cow to the date of next calving is called ....
11. .... and .... are the methods of identification of animals.
12. Main two types of milking are .... and ....
13. Gestation period of Gir cow is.........
14. Milk content .... % water.
15. Murrah is breed of ....
16. Tagging means..........
17. The frozen semen straws are stored at ....^{0}C temparature.

18. Dry period of cattle is .... days.
19. Feeding standards are of .... types.
20. Semen straws are stored in .... at -196 ^{0}C temp.

Multiple Choice Questions

1. Female animal that have not been bred is called as ....
   a) Open animal b) Stud animal
   c) heifer d) Cow
2. Hump is present in ....cattle.
   a) Indian cattle b) Buffaloes
   c) crossbred cows d) exotic
3. Milk replacer is fed in ....form to calves.
   a) Solid b) liquid
   c) gruel d) pellet
4. .... is the bull whose ancestral record is known.
   a) Pedigree bull b) bull
   c) Teasure bull d) stud bull
5. Tightly curved horns is the feature of ....breed.
   a) Murrah b) Jersy
   c) Mehsana d) Surti
6. Separation of calf from dam is called as..........
   a) Casting b) weaning
   c) drying off d) calving
7. First time deworming in calf should be carried out at .... days of age
   a) 8 b) 15
   c) 30 d) 90
8. A male calf under one year age is called as............
   a) Bull calf b) bull
   c) breeding bull d) heifer
9. Letting down of milk is affected by........
   a) Oxytocin b) growth hormone
   c) Estrogen d) L.H.

10. Milk feeding rate in second month of age is …….% of body weight.
    a) 10 b) 15
    c) 5 d) 20
11. Best method of hand milking………….
    a) Stripping b) None of these
    c) knuckling d) full hand
12. General dairy farm practices includes…………….
    a) Tagging b) castration
    c) weighing d) all of these
13. In isolation box each adult animal requires…….sq.ft. area/floor space.
    a) 100 sq. ft. b) 200 sq.ft.
    c) 150 sq.ft. d) 100 m.sq.
14. Roughages are of ………types
    a) Two b) Three
    c) four d) Six
15. Cereal fodders are rich in……………
    a) Carbohydrates b) protein
    c) crude fibre d) fat
16. Milk mirror of cow is……….
    a) Udder size b) milk vein
    c) abdominal size d) teat size
17. Common infectious diseases of cattle & buffaloes are……….
    a) H.S. B) B.Q.
    c) F.M.D. d) all of these
18. Best buffalo breed in India for milk production is………
    a) Pandharpuri b) Surti
    c) Jaffrabadi d) Murrah
19. Feeds that contain less than 18% crude fibre and more than 60% TDN are called ………… such as grains, oilcakes, grain by products etc.
    a) Roughage b) feed suppliments
    c) feed additives d) concentrates

20. The process of including certain animals in a population for becoming parents of next generation

    a) Breeding b) selection
    c) insemination d) casting

21. Best draught breed of India is.........................

    a) Nagori b) Bachur
    c) Kankrej d) Amritmahal

22. The colostrum should be fed to calf within....... of birth.

    a) 12 hrs. b) 6 hrs.
    c) 2 hrs. d) 24 hrs.

23. Milk feeding rate in first month of age is ....% of body weight..

    a) 5% b) 10%
    c) 15% d) 20%

24. .... is the first solid feed given to calves?

    a) Milk replacer b) Calf starter
    c) Calf ration d) None of these

25. House of cattle and buffalo is called.....

    a) Shed b) Byre
    c) Barn d) All of these

26. Letting down of milk is affected by.........

    a) Oxytocin b) Estrogen
    c) LH d) Progesteron

27. Sickle shaped horns is the feature of ....breed

    a) Murrah b) Mehsana
    c) Surti d) Jaffrabadi

28. Act of parturition in cattle is called as...........

    a) Lambing b) Kidding
    c) Foaling d) Calving

29. A female calf under one year age is called as............

    a) Calf b) bull calf
    c) heifer d) bull

30. .... is the synoname of Sahiwal cattle

    a) Lola b) Dongarpatti
    c) Nellore d) Surti

31. First time deworming in calf should be carried out at .... days of age
    a) 80 days
    b) 30 days
    c) 8 days
    d) None of these
32. Isolation box is provided for .....
    a) ill animal
    b) weak animal
    c) animal suffering from contagious disease
    d) All of these
33. A system of breeding between very closely related animals is called as............
    a) Outcrossing
    b) inbreeding
    c) cross breeding
    d) criss-crossing
34. Canine/incisor teeths present in ————jaw of cattle
    a) lower
    b) upper
    c) Both lower & upper
    d) None of them
35. Calving box is provided for .....
    a) Non-pregnant animals
    b) calves
    c) ill animals
    d) pregnant animals
36. Concentrates are of .... types
    a) Three
    b) One
    c) Ten
    d) Two
37. '8' knot is used for controlling the cow while...........
    a) Milking
    b) Treatment
    c) Deworming
    d) Throwing animal on ground
38. Leguminous fodders are rich in..........
    a) Carbohydrate
    b) Protein
    c) Fat
    d) None of these
39. .... method of milking should be avoided as it is painful for cow and may injure udder and teats.
    a) Stripping
    b) Knuckling
    c) Machine milking
    d) Full hand milking
40. Calf starter is fed to calves in .... form
    a) Solid
    b) liquid form
    c) Gruel
    d) pellet form

41. For advance pregnant animals, in addition to maintenance ration, additional amount of ———— kg. Concentrate mixture is recommended for zebu cows and 1.75-2.5 kg for crossbred cows, buffaloes.

    a) 1.25-1.5 kg   b) 2.0-3.0 kg
    c) 0.5-1.0 kg   d) Nil

42. The time interval between date of drying off the cow to the date of next calving is called as........

    a) Service period   b) Gestation period
    c) Dry period   d) none of them

43. .... is the System of breeding in which pure bulls are used for improvement in non- descript females for several generations.

    a) Back crossing   b) Grading up
    c) Out-crossing   d) Cross breeding

44. The total amount of feed that an animal is offered during a 24 hour period of time is called as ....

    a) Roughages   b) Feed
    c) Bulk   d) ration

45. Systems of breeding are of .... types

    a) Two   b) One
    c) Six   d) Nine

46. Roughages are of .... types

    a) Three   b) One
    c) Ten   d) Two

47. Bull nose ring is used for controlling the ....

    a) Bullock   b) Milking cow
    c) Bull   d) Throwing animal on ground

48. Cereal fodders are rich in..........

    a) Carbohydrate   b) Protein
    c) Fat   d) None of these

49. SNF in milk of cow is depends on

    a) Milk protein content   b) Milk mineral content
    c) Milk lactose content   d) All of the these

50. Mehsana is the cross between ....&................. buffalo
    a) Surti & Jaffrabadi
    b) Murrah & Surti
    c) Murrah & Nagpuri
    d) None of these
51. Milk yield of buffalo depends on———
    a) Breed
    b) Feeding management
    c) health care
    d) All of these.
52. Milch purpose breed of India is.........
    a) Nagori
    b) Deoni
    c) Amritmahal
    d) Sahiwal
53. Youngone of cow is called as———
    a) Bullock
    b) Calf
    c) Heifer
    d) Cow
54. Milk fat in milk means—
    a) Protein in milk
    b) SNF in milk
    c) Sugar in milk
    d) None of these

Define the following

1. Dehorning
2. Castration
3. Artificial insemination
4. Service period
5. Grading up
6. Breed
7. Embryo transfer
8. Bull
9. Casting
10. Grooming of animals
11. Anand pattern
12. Knuckling
13. Selection
14. Pedigree bull
15. Parturition
16. Dehorning
17. Gestation
18. Bullock
19. Ration
20. Stripping
21. Animal husbandry
22. Cross breeding
23. disbudding
24. Dry period
25. Progeny testing
26. Management
27. Concentrate
28. Roughage
29. Balanced ration
30. Heterosis

31. Criss-srossing
32. Breeding
33. Fertilization
34. Heat synchronization
35. Culling
36. Judging
37. Nutrients
38. Feed suppliment
39. Feed additives
40. Back crossing
41. In breeding
42. Oat breeding
43. Estrus
44. Calving
45. Lactation length
46. Energy
47. Fisting
48. Milking
49. Let down of milk
50. Feeding standards
51. Ovulation
52. Restraining of animals
53. Maintenance ration
54. Gestational ration
55. Production ration

Answer the following

1. Which breed of buffalo is evolved due to crossing of Murrah & Surti breeds?
2. Milk production of animal depends upon which characters?
3. What is the total milk production of India in 2017-18?
4. Which factors affect Milk fat percentage?
5. What is the meaning of SNF in milk?
6. What is the life span of cattle?
7. Enlist the parts of female reproductive system of cow.
8. Enlist the methods of casting
9. Define concentrate
10. Write the examples of green roughages
11. Write the Classification of feedstuffs
12. Define casting
13. Enlist the milch purpose breeds of India
14. Write the Synoname of Sahiwal breed
15. Oily skin is found in which breed?
16. Mehsana is the cross between which breeds?
17. Best draft purpose cattle breed in India
18. Write the examples of leguminous fodders

19. Write the floor space requirement of calves upto 6 month age in covered area.
20. Define production ration

State whether True or False. If false, rewrite the statement after making necessary corrections

1. Milk replacers can be started from 2nd week of age.
2. Dr. V. Kurian is the father of white revolution.
3. Cattle are non-ruminant animal.
4. Anand pattern is of two tier dairy cooperative structure.
5. Deoni is dual purpose breed.
6. Tagging is the method of identification of animals.
7. Colostrum should be fed to calves @10% of the body weight.
8. Sahiwal is draft purpose breed of cattle.
9. Murrah is highest milk yielder amoungst the buffalo breeds.
10. Knuckling is safest method of milking.
11. Deoni is milch purpose breed.
12. Average Milk fat content of buffalo milk is 3.5-4%
13. Bull is young one of cattle.
14. Liver is the oragn of digestive system of cow.
15. Milk replacer is content less protein than concentrate ration for adults.
16. Artificial insemination means use of artificial semen for service of cow or buffalo.
17. Only maintaining the Milk yield record is more helpful for deciding the farm is in profit or loss.
18. Requirement of Dry fodder is more than green fodder in ruminants.
19. Loose housing system is very costly than other systems of housing.
20. Kadbi is concentrate.
21. Growth hormone is required for milk let down.
22. Colostrum should feed to calf after 2 hours of parturition.
23. Quality of milk production is only depends upon breed of cattle.
24. Loose housing system is very laborious method of housing.
25. Foot and mouth disease can be prevented by vaccination against this disease.

Match the pairs

1.

A	B
1. Ear tagging	a) Identification
2. Hot iron	b) Dehorning
3. Burdizzos castrator	c) Castration
4. Fisting	d) Milking
5. Financial register	e) Dairy form record

2.

A	B
1. Natural service	a) Bull
2. Artificial insemination	b) Frozen semen
3. Oxytocin	c) Let down of milk
4. Adrenaline	d) Milk hold in udder
5. Colostrum	e) Immunity development

3.

A	B
1. Lola	Loose skin/Sahiwal
2. Oily skin	Dangi cattle
3. Jet black colour	Murrah buffalo
4. Sickle shaped horns	Surti buffalo
5. Tightly curved horns	Murrah buffalo
6. Swai chal	Kankrej cattle
7. Copper skin colour	Bhadwari buffalo
8. Milch breed of cattle	Gir, Sahiwal, Red sindhi, Tharparkar
9. High fat percentage in milk	Surti buffalo
10. Highest milk producing cow	Holstein Friesian cow
11. Yellow tinged ears	Tharparkar cattle
12. Sleepy look/almond shaped eyes	Gir cow
13. Double dished forehead	Jersey cattle
14. Dry roughage	Kadbi
15. Leguminous fodders	Lucerne, berseem, cowpea
16. Cereal fodders	Maize, Jowar, Hybrid napier varieties
17. Perennial cereal grass	Hybrid napier varieties
18. Perennial leguminous fodder	Lucerne, Dashrath

19. Silage	Fermented fodder
20. Hay	Dried fodder with high nutritive value
21. Haylage	Low moisture silage
22. Kadbi /straw	Dried fodder with low nutritive value
23. Energy rich concentrate	Maize, Jowar, Bajra grains
24. Animal origin protein concentrate	Meat meal, fish meal, blood meal
25. Plant origin protein concentrate	Oilseed cakes, chunies, Dals
26. Feed supplement	Mineral and vitamin supplement
27. Feed additives	Antibiotic, probiotic, prebiotic, Antioxidants
28. Testicles	Sperm production, testosterone secretion
29. Ovary	Ova production, LH and FSH production
30. Fertilization	Infundibulum
31. Gestation period of cow	282 days
32. Gestation period of buffalo	310 days
33. Dry period	60-90 days
34. Service period	60-90 days
35. Intercalving period in cow	365 days
36. Intercalving period of buffalo	420 days
37. Total dry matter requirement for indigenous dry cow	2% of body weight
38. Total dry matter requirement for indigenous milch cow	2.5% of body weight
39. Total dry matter requirement for Buffaloes and crossbred dry cow	2.5% of body weight
40. Total dry matter requirement for Buffaloes and crossbred milch cow	3% of body weight
41. Milk replacer	20-28% CP
42. Calf starter	18-20% CP
43. First solid feed for calf	Calf starter
44. Feed additives	Non-nutritive in nature

45. Feed supplements	Nutritive in nature
46. Young bull ration	DCP=12-15%
47. Ration for Bull in service	DCP=10-12%
48. 1 kilocalorie	1000 calories
49. 1 megacalorie	1000 Kcalories
50. 1 calorie	4.185 Joule
51. Inbreeding	Breeding of related animals
52. Inbreeding	Close and line breeding
53. Outbreeding	Breeding of unrelated animals
54. Outbreeding	Outcrossing, cross-breeding, Species Hybridiazation and Grading up
55. Cross-breeding	Criss-srossing, Triple cross and back cross
56. Corpus luteum	Secretion of progesterone
57. Diestrum	Longest phase of estrus cycle
58. Estrus	Period of receptivity or sexual desire by female
59. Pedigree selection	Selection on the basis of ancestor performance
60. Progeny testing	Selection on the basis of progeny performance